Environmental Restoration in the Era of Frontier Science

Proceedings of the Third International Conference on Science and God

Global Online Conference
April 11-12, 2022

Hyo Jeong International Foundation
For the Unity of the Sciences Press

This Proceedings volume contains content delivered at the
Third International Conference on Science and God (ICSG III).
This book is the Second Edition.

Book design by Gil Roschuni
Front cover photo by Nikola Majksner/unsplash
Back cover photo by Dương-Nhân/pexels

Library of Congress Control Number: 2023909354
ISBN: 978-1-947499-12-6

Published by
Hyo Jeong International Foundation for the Unity of the Sciences Press

3600 New York Ave. NE
Washington DC 20002
USA

324-211 Misari-ro, Seorak-myeon
Gapyeong-gun, Gyeonggi-do
Republic of Korea

Table of Contents

Preface

The Third International Conference on Science and God (ICSG III) was held on April 11–12, 2022 (Korea time), using a virtual platform covering the globe. The overall theme for the conference was, "Environmental Restoration in the Era of Frontier Science." This book presents the proceedings of this conference.

ICSG III was a landmark event. The presentations demonstrated that we have entered the era of postmaterialist science, while also looking into new developments that offer ways to resolve environmental challenges. The conference was sponsored by the Hyo Jeong International Foundation for the Unity of the Sciences (HJIFUS), as a joint project of HJIFUS offices in the United States and Korea, with support from related offices in Japan and Europe.

HJIFUS was founded in 2017, based on the vision and principles taught by Rev. Dr. Sun Myung Moon and Rev. Dr. Hak Ja Han Moon. It has the mission of identifying, promoting, and implementing the most effective solutions to environmental challenges. As a foundation for fulfilling this mission, HJIFUS has been sponsoring the annual conference series known as the International Conference on the Unity of the Sciences (ICUS). This series has been focusing on exploring technologies derived from conventional scientific approaches to help resolve environmental issues. In 2020, HJIFUS launched the ICSG series to broaden our search by considering new approaches based on frontier science.

Following the pattern of ICSG II last year, ICSG III was also convened online. The COVID-19 pandemic showed no signs of abating, and we had to rule out an in-person meeting. However, advances in electronic communications technology allowed us to run the conference using a virtual platform that connected participants across multiple time zones around the world. For this, we engaged the services of iPeaceTV in Korea, which has state-of-the-art equipment and a highly skilled staff.

ICSG III had 72 registered attendees representing 8 countries, including researchers who were willing to expand the scope of their investigations outside the mainstream box. The structure of the conference was essentially the same as that of ICSG II: an opening session, including a keynote address, followed by three academic sessions in plenary format. Each academic session consisted of presentations on two topics, with one presenter and two commentators per topic, and a discussion. The conference ended with a closing session in which the session chairs gave summaries of the presentations and discussions at their respective sessions.

Through ICSG I and ICSG II, we opened the door to frontier science research, and we looked at the beginnings of the postmaterialist era. Now, through ICSG III, we could see some of the features of the postmaterialist era, along with the results of research that demonstrate the need for scientists to shift their paradigm from a materialist to a postmaterialist one. In addition, there was a presentation on energy medicine, which offers a way to heal illnesses caused by environmental pollution, and another talk compared a variety of energy-producing technologies that can avoid the production of greenhouse gases. Through the eyes of quantum physics, as well as spiritual practices, we humans are interconnected with all things in nature. Therefore, by treating the things of nature with care and by living in harmony with it, we can create a beautiful living environment.

In the Appendix of this book, one can find information about the conference organizing committee and list of attendees, as well as some background on the founders of ICSG and HJIFUS. There is also a short introduction to HJIFUS, along with the boards of directors and contact information for our worldwide offices.

I wish to thank all those who contributed to the success of ICSG III, including the session chairs, presenters, commentators, and other attendees from multiple time zones. I also thank the staff who helped with prerecording the speeches, and everyone who invested their efforts to produce this volume. ◊

Dr. Douglas D.M. Joo
Chairman, ICSG III and HJIFUS

ICSG III

OPENING SESSION

Environmental Restoration in the Era of Frontier Science

Dr. Douglas D.M. Joo, Chair of ICSG III, delivers the Welcoming Remarks.

Dr. Sun Jin Moon gives the Founder's Address on behalf of Rev. Dr. Hak Ja Han Moon, Cofounder of ICSG.

Dr. Lisa Miller gives the Keynote Address.

ICSG III participants tune in for the Opening Session.

Opening Session

Welcoming Remarks

Dr. Douglas D.M. Joo
Chair, ICSG III

Respected Scientists, Scholars, Ladies and Gentlemen:

On behalf of the Hyo Jeong International Foundation for the Unity of the Sciences (HJIFUS), I am very happy and honored to welcome you to the Third International Conference on Science and God (or ICSG III). This year, the conference theme is, *Environmental Restoration in the Era of Frontier Science.*

As you may know, HJIFUS is dedicated to finding, promoting, and implementing the most promising solutions to environmental problems. Toward this end, we have been sponsoring a conference series called the International Conference on the Unity of the Sciences (ICUS), held annually over the past five years.

I should note that the ICUS series was founded in 1972 by Rev. Dr. Sun Myung Moon and Rev. Dr. Hak Ja Han Moon. At that time, the underlying motif was to promote unity among the sciences, guided by absolute values, that is, values that benefit all humanity. However, with the founding of HJIFUS in 2017, ICUS meetings have been focusing on addressing environmental issues, mainly through conventional scientific approaches. In 2020, we launched the first ICSG meeting in Seoul, Korea, thereby opening the door to unconventional, "frontier science" approaches to broaden our search for the most promising solutions to environmental problems.

You may ask why we use the term "Science and God" in the name of our new conference series. Allow me to explain that it is not intended to indicate that the ICSG meetings will blend science with a particular theology. Rather, it implies a flow in the history of Western scientific thought: from a classical, materialist perspective to the revolutionary ideas of a postmaterialist viewpoint. As frontier science investigates the mysteries of consciousness and other phenomena, it may lead to a better understanding of unseen dimensions, possibly revealing fundamental approaches to resolve environmental challenges. Ultimately, these investigations may lead to an understanding that the First Cause may have attributes of intellect, emotion, and will, thereby making it possible for humans to communicate with the First Cause and discover more complete ways to address environmental issues.

The ongoing COVID-19 pandemic and environmental issues such as climate change have one thing in common: they are global in scope. They point to the fact that all people are interconnected with one another and with nature. Therefore, to tackle these problems, the people of the world need to work together, cooperatively. This is also a reason why these conferences are held on a global scale, and technological solutions need to be shared with all people in all nations. In this manner, we need to follow the principles of interdependence, mutual prosperity, and universal values, which will lead to world peace.

Last year, on account of travel restrictions imposed by the pandemic, we held ICSG II as an online meeting. The presentations made at ICSG I and ICSG II showed us that science has moved into the postmaterialist era. Now, ICSG III is being held on the foundation laid by the two preceding conferences in this series.

Once again, for this year's meeting, we are using a virtual platform to circumvent the pandemic-related restrictions in international travel. I wish to congratulate and thank all of you, distinguished guests, for overcoming the difficulties imposed by the pandemic, including the need to transcend differences in time zones, to attend this conference. At the same time, I applaud you for engaging in pioneering research that goes beyond the boundaries of conventional science.

ICSG III will cover a range of thought-provoking presentations by scientists and scholars who have the courage to step outside the box of traditional science to explore the deeper dimensions of reality. For example, Prof. Lisa Miller, our keynote speaker, will tell us how science is awakening us to our inner, spiritual dimension.

After that, the first main session will examine the characteristics of a postmaterialist era and phenomena that require a shift in the scientific paradigm from a materialist to a postmaterialist perspective. The second session will look at applications of frontier science in new modes of healing and novel technologies to generate energy while minimizing the release of greenhouse gases. The third session will probe the connections between people and the natural world, and envision the Earth's future based on living in harmony with nature. In this manner, we hope that this conference will show us how to change human consciousness and develop a wholesome attitude toward nature.

A few days from today, on April 16, is the 62nd holy wedding anniversary of our Founders. Please join me in congratulating them and expressing our gratitude for their untiring efforts to protect the environment and to enhance the quality of life for all humanity.

Another significant day this month is April 22, which is observed as Earth Day. I am pleased to announce that one year ago, HJIFUS launched an online, bimonthly publication, *The Earth & I.* Its purpose is to educate the general public on environmental issues, and to inspire people to take an active role in caring for the environment.

Before I close, I wish to offer a salute to our esteemed friend, Professor Luc Montagnier, who passed away on February 8 this year. He received the 2008 Nobel Prize in Physiology or Medicine for his work in discovering and studying HIV, the virus that causes AIDS. He contributed to both our conference series, ICUS and ICSG, as a speaker and as a co-chair.

Finally, I wish to thank the iPeaceTV technical team and time zone coordinators for their tremendous investment in setting up and operating the virtual platform, which connects everyone together, as if we were in the same location. In addition, I am grateful for the support we have received from The Washington Times Foundation in the United States and *The Segye Times* in Korea.

The success of this conference can be measured in terms of how much it can help change human consciousness and attitude in order to protect and restore the natural environment. Once again, I thank you all for your participation.

Thank you very much. ◊

About the Speaker

- Chair, Hyo Jeong International Foundation for the Unity of the Sciences
- Former Chair and President of *The Washington Times* and United Press International, and several other international news publications under News World Communications, as well as AmericanLife TV Network and other electronic media
- Former President, World Media Association, The Washington Times Foundation, American Family Coalition, and Times Aviation International
- Former International President of the Youth Federation for World Peace
- Former Co-Chair of the Board of Trustees of the University of Bridgeport, Connecticut
- An honorary doctorate in Political Science from Sun Moon University; M.Ph. in Political Science, George Washington University

Opening Session

Founder's Address

Dr. Sun Jin Moon
Representing the Founders

Distinguished Participants attending this Third International Conference on Science and God (ICSG III):

Greetings! I am Sun Jin Moon, daughter of Dr. Hak Ja Han Moon, Founder of the ICSG series, and I am truly excited to see you all. You are among the greatest scientists, thinkers, and illuminating minds from across the globe, and it is indeed my great honor to welcome you and share my parent's insights and vision.

I am especially pleased to acknowledge the participation of Professor Lisa Miller as our keynote speaker. I have had the greatest privilege to learn from her at Columbia University Teachers College, where she has established the Spirituality Mind Body Institute. She is the vanguard of enlightenment in the science of spirituality and is trailblazing the holistic path for optimal health and well-being. She is a true advocate for all life—and the whole of creation. Thank you, Dr. Miller, for your amazing work and your presence here!

Today, we are gathering virtually from multiple time zones around the world, so I wish all in attendance a glorious morning, a lovely afternoon, and wonderful evening. Wherever you are Zooming in from, we thank you for your participation, passion, concern, and care for the future of our planet. Through these integrative and collective exchanges, we hope to discover and chart the paths towards peace and prosperity for all people and all life on the planet. This noble goal of world peace has been the mission of my parents, who have devoted their lives to this great cause. I am deeply moved to witness this third ICSG, where we are all wholeheartedly called to pave a future path toward that dream for a world of harmony and cooperation.

The theme for today's conference is, *Environmental Restoration in the Era of Frontier Science,* and rightly so, since the world we see around us is in deep need of healing and restoration. The word "restoration" means the act of correcting, reforming, or returning something to its original condition, unimpaired or improved (*Merriam-Webster's Dictionary,* no date). This restoration is needed not only for nature and this planet, but also for an even greater and primary task: It needs to be restored in the hearts, bodies, minds, and souls of all Divine life on it. This is our mission in life, to restore our internal and external nature back to its original form or original mind, thereby manifesting true love, Divine and collective oneness, harmony, joy, and peace.

Now more than ever, peace and prosperity for all seems like a distant dream. We are all interconnected and interdependent, and the threat to peace anywhere is a threat to peace everywhere. We are confronted with the threat of North Korean missiles in the East, while at the same time we are watching the horrors of Russia's war in Ukraine that threatens the West. No place on earth is free from the consequences of these conflicts, as underscored by multilayered economic uncertainty.

At the same time, the ongoing COVID-19 pandemic and the emergence of new viral variants have us in a race for new vaccines. As cases spike and fall, we are left in a dizzying battle to find some normalcy and stability in our home, family, work, and life.

We are starting to realize that nothing is stable, as the planet is reeling under the effects of climate change, severe weather, extreme drought, record-breaking fires, floods, storms, and tornados carving up and destroying our beloved hometowns. We are witnessing the displacement of the world's peoples on a daily basis, as well as the mounting loss of precious species on earth. As all the glorious, life-sustaining gifts from the universe are being polluted, squandered, and exhausted, we are witnessing more people in need; as a result, the humanitarian and environmental aid needed around the globe is skyrocketing everywhere. We are in constant battle with each other, divided among ourselves, and all life on this planet is suffering.

In February 2017, when addressing scientists at the Twenty-Third International Conference on the Unity of the Sciences (ICUS XXIII) in Seoul, Korea, my mother acknowledged the material benefits of scientific research. Yet she also saw the problems that have arisen from a selfish mindset. She told them:

> *You should know about God.... We are moving in a direction that threatens the lives of human beings and all creatures, even to the extent that there may be no future for our Earth and our race. We cannot ignore this reality. We cannot think about only what we need now, today.*

Time is not on our side. With an ever-dwindling supply of sustenance to live on and a compounding need to consume what is left, we have tipped the scales of the Earth so far that we are at an existential crisis.

So, where is the path for peace in these unsustainable times? We seem to be losing every gift of pristine and abundant creation due to human greed and heartlessness. How can we restore this world, and the beauty and amazing gift of creation, if we cannot restore our own hearts so that we live in gratitude, stewardship, responsibility, harmony, and love for all life? The answer to this question was offered by my father, Rev. Dr. Sun Myung Moon, in his autobiography titled *As a Peace-Loving Global Citizen* (2010, pp. 38–39):

> *Even the smallest grain of sand contains the principles of the world, and even a speck of dust floating in the air contains the harmony of the universe. Everything around us was given birth through a combination of forces so complex we cannot even imagine it. These forces are closely related to each other. Nothing in the universe was conceived outside the heart of God. The movement of just one leaf holds within it the breathing of the universe.... We need to feel the difference between the sound of spring rain falling like a soft whisper and that of the autumn rain falling with pops and crackles. It is only the person who enjoys resonance with*

nature who can be said to have true character. A dandelion blooming by the side of the road is more precious than all the gold in the world. We need to have a heart that knows how to love nature and love people. Anyone who cannot love nature or love people is not capable of loving God. Everything in creation embodies God at the level of symbol, and human beings are substantial beings created in the image of God. Only a person who can love nature can love God.

The key to return and restore our original mind, body, and soul to resonate with the highest consciousness and Divine Being from the inception of all life in the cosmos is love. We must know how to "love nature and love people." Then we will know our true internal nature which is imbued with Divine love, happiness, oneness, and peace with all life and creation. By understanding the Divine heart of God, Allah, Yahweh, Atman, higher power, spirituality or whatever resonates with you, it is that give and take, deep resonance, and symbiotic relationship of a collective consciousness that opens our hearts, minds, bodies, and souls to revere, care, honor, and love this life and all creation. It is only when we tap into this inner power of unconditional love that the external result will be a world we all dream of, a world of peace, wellness, and joy.

Today, I am proud to be among the great minds and hearts that are bringing this theme, *Environmental Restoration in the Era of Frontier Science,* to life. I hope we can all return to the highest source of wisdom that leads us to this dream of peace, prosperity, and well-being for people and the planet. Thanks to scientific development, we can now hope to unveil the world that lies beyond matter and even encounter God, the ultimate cause of the universe. My expectation is that ICSG will explore what has not been touched on by traditional science and play a meaningful role in diffusing myths about the relationships between the environment and human beings, between mind and body, psychology, and physiology, and spirituality and so on. By doing so, ICSG will contribute to the opening of a new history of science and enhance the ways by which we can restore the environment.

I thank you all for your great contributions, and I know that your work will change the trajectory of this world from existential destruction to enlightened restoration, reconciliation, sustainability, and peace for future generations to come. I hope this conference, anchored in a shared sense of solidarity, will prove a valuable opportunity for each of you to offer your ideas and research results to change this world toward a higher, peaceful reality. I look forward to hearing your presentations and deliberations during this conference.

Thank you very much. ◊

About the Speaker

- Fifth daughter of Rev. Dr. Sun Myung Moon and Rev. Dr. Hak Ja Han Moon
- Senior Vice President, Women's Federation for World Peace International
- Chair, Pacific Rim Education Foundation in Kailua-Kona, Hawaii (since 2012)
- Has served the Family Federation for World Peace and Unification, Universal Peace Federation, and Women's Federation for World Peace (since 2000)
- Baccalaureate in psychology from Harvard University
- Inducted into the Alpha Sigma Lambda National Honor Society
- Candidate for Master of Arts degree in Clinical and Counseling Psychology from Columbia University
- Received an honorary doctorate in Social Science from Sun Moon University

Opening Session | Keynote Address

Exploring the Unknown: Where Is Science Leading Us?

Professor Lisa Jane Miller
Psychology and Education,
Teachers College, Columbia University
New York, USA

I foremost want to thank the organizers of this global conference. It is a great honor and a joy to be with you.

I am sharing today a body of science that was not on the map in aggregate 20 or 30 years ago, but in the past 10–15 years has burgeoned in such a way as to carry deep implications about who we are as human beings, and what our core purpose and possibilities are here on Earth.

I will start by sharing a story that pertains to a child because, as many of you may realize, we come into this life fully aware, born as naturally spiritual beings. This is a story of my beloved son, Isaiah, whom I adopted at about 10 months from a small orphanage thousands of miles from our home.

When Isaiah was about two-and-a-half years old, he and I were in the park, enjoying the sunny day—the glistening on the water, and the green grass. Suddenly, we spotted yellow, fluffy, baby geese (**Figure 1**). Isaiah thought this was wonderful, so he ran up and wanted to be friends with the geese. With great enthusiasm and joy, he started waving and padding toward them. He did this with his whole heart, but it was very frightening to Mother Goose. When she saw this robust human coming toward her little children, she turned with a huge protective hiss—but she did not turn toward Isaiah, she turned and hissed at me.

Figure 1

Mother Goose said to me what any mother of any species on Earth might say, which is, "You have not been watching over your child to the protection of mine." Now, Mother Goose was in relationship with me. She knew Isaiah was just a baby; she knew I was the mother; she knew we had like responsibilities.

All living beings on Earth, are knowers and in relationship with one another. However, we have created a culture around the world in which we have all but gouged out our eyes, unable to see the reality of relationships all around us.

So, what does that mean? What does it mean for the child, for the mother, and for Mother Goose? It means that we live in a situation where humans must be very scary. There is a symphony of life in which the geese, the fish, the birds, and everyone else are in relationships. However, when we come lumbering around—basically anyone over the age of three—we are dangerous because we are not moving with the rhythm of the universe. We are not aware of the many relationships that are in action right before us, de facto. Yet Isaiah, like any young child, was intimately aware of this foundational relationship.

Figure 2

Isaiah's story is the story of any child and all of us here on Earth. It has to do with our natural endowment to be in relationship with our fellow living beings and with the deeper force, the deeper presence—what I call God, and some call Spirit—in nature.

Science Pointing to Spiritual Awareness

We now have very good, peer-reviewed science published in mainstream medical journals—such as the *American Journal of Psychiatry* and the *Journal of the American Medical Association*—and supported by the American Academy of Child and Adolescent Psychiatry, pointing to this innate human endowment. I call this endowment our "spiritual awareness."

Across the many beautiful faith traditions on Earth, we have thousands of ways to connect into the deep spirit, to talk to God, Spirit, the universe, Hashem, Allah, Jesus, whatever one's word may be. These are rich faith traditions, our religions, our wisdom traditions, our cultural traditions. For about two-thirds of people, their faith tradition or wisdom tradition goes hand-in-hand with their natural spiritual awareness.

For about one-third of people on Earth, they will say, "Yes, I indeed do feel connected, I am a spiritual being, but I do not share in a faith tradition or a wisdom tradition." However, whether or not we are part of a religion or tradition, each of us is born with this natural capacity, represented by the yellow circle in **Figure 2**.

Since there is a rich array of expressions and symbols, we might say there is one Source of life—one mover, one spirit, a loving guide, an intentional spirit through life—one Source with many names. As there is one Source in and through all of us, it leads us to feel and know a deep love of neighbor, and we might have a deep sense of awareness in the presence of one another.

We might feel and see a deep unitive reality in each other's presence. For example, we may wonder, "Why have you come right here, right now? Why do we have this moment together, at this instant, so that of all people on Earth I can speak to you?"

When we know that there is a unitive reality and connection with the Source of life, the Higher Power, and when we know that the Higher Power exists through every one of us, then we no longer look upon each other as merely the bio-body suit, the way we are zipped up and dressed up as souls on Earth. We are able to feel and see into one another the same presence that my son felt in seeing the baby geese.

Figure 3 gives a road map of the science of spiritual development. It is a blueprint of our human composition as knowers, as spiritually aware, innately spiritual beings.

The first of the four axioms of science is that every one of us is born on Day One as a spiritual being. How do we know this? Science uses a "twin study." We look at twins raised together and twins raised apart, and we factor out the degree of commonality as a function of shared genes and shared environment.

We have an innate capacity through which we experience a relationship to the Higher Power and feel the existence of Spirit through one another. By the time we are 30 years old, however, we say it is about one-third innate and two-thirds environmentally socialized.

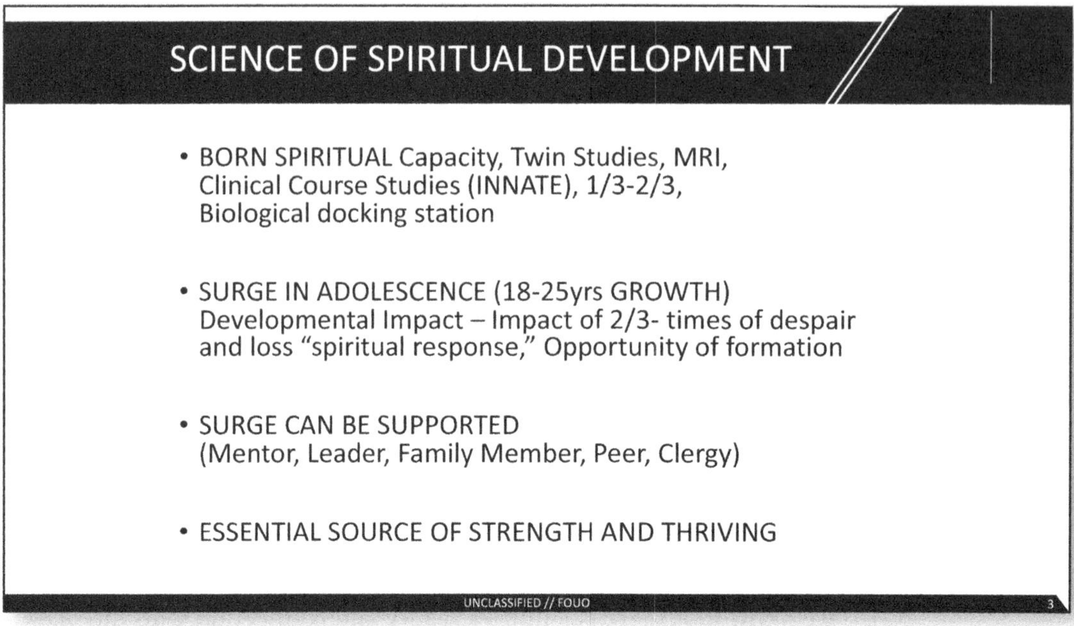

Figure 3

Let us go back to Isaiah. When he was two-and-a-half, there had been very little external socialization. He spoke from his natural awareness.

However, by the time we are 30, we have received many socializing messages—that animals are somehow different from us, that, "Oh, that little duck, that little goose, he just wants your food." The relationship is seen as instrumental or transactional, instead of one of common heart, which is a transformational relationship. The embrace of environmental socialization is very powerful, and it can be so severe that it blocks us from seeing through our naturally endowed eyes.

Seeking Our Life's Purpose

Nonetheless, there is great hope. Just as we are hardwired to connect, so too we are hardwired for renewal. With adolescence, puberty, and coming-of-age, physically, there is a burgeoning of our spiritual capacity.

From the inside out, we hunger to connect to all living beings and to the source of life. We, in our heads, long to understand life's ultimate purpose. We wonder, "What is the purpose of life itself? Is there a Higher Power?" These are universal longings.

Search Institute researcher Peter Benson and colleagues looked at youth populations in Thailand, Ukraine, Australia, Canada, the U.S., and several other countries. They found that no matter where a teen is growing up, he or she hungers to understand the nature of reality and wants their actions—how they use their own precious life—to be of consequence to the greater realization of goodness on Earth. That is the deep, underlying intention built into our lives.

So, this process, this burgeoning, can be harnessed in education, in universities and schools, and again at midlife. What we call a "midlife crisis" is not just a time when people buy new sports cars or think they need a better house or a new spouse. A midlife crisis is another hardwired booting-up, when we hunger to see more deeply into life, to live in a way that is in keeping with the most profound possibilities, opportunities, and intentions of the universe.

When we support one another—our children, colleagues, and friends—in realizing our spiritual nature, and when we walk in such a way as to speak truth from our deep hearts, we are providing the two-thirds environmental socialization by which we can help one another realize our natural, spiritual potential. This has often not been the case, and we needed to make a change yesterday, or at the very latest, now.

Postindustrial Culture

There is enormous suffering in postindustrial culture, because we have acted as if we are pieces and parts of a machine, one lever on an assembly line to be pulled, or one cog in a watch to turn. However, we are not pieces and parts when seen through the eyes of the consciousness-based sciences. We are more aptly understood as whitecaps of waves on an ocean. We are part of one body of life that is seen in the moment through which we are connected with one another.

The view that we are nothing but parts of a machine leads us to wonder:

"How good a cog am I? How well do I pull the lever?" This notion is a latent, atomistic view that we are alone and that we are merely functional beings that exist to produce.

At the moment, across the world, in postindustrial countries, there is a rate of addiction, a rate of depression, a rate of suicide in young adults that surpasses the rate of death by auto accidents. There is suffering, and it is not because of the state of the world. It is not because of the geese, not because of the tragic conflicts, and not because of difficulties in the economy.

The suffering around the world is because of our inner seat of awareness. Postindustrial culture has told us that we are alone, that we are valuable only insofar as we produce. We are using what I call only half the brain—half the brain of achievement awareness. We may think, "I am better if I do more, if I earn more money, if I am more famous, if I take on a bigger role." However, achievement awareness is very brittle, because sooner or later there comes a day when each of us faces someone who is bigger or better. Also, achievement awareness is a very thin slice of how we function in life. It says that if we get A plus B plus C all lined up, we will get to that red door. That promotion is yours; that marriage is yours; that new house is yours. However, as we reach for that red door to turn the handle, we may find that it is stuck.

Discovering the Yellow Doors

There is no time where that has been more apparent than during the COVID pandemic. "Suddenly, everything I planned for is stuck: the promotion is not mine; they hired someone else; they brought in somebody." Such moments, when the red door is stuck, can lead to frustration, anger, and despair.

Yet only because that red door is stuck do we have the opportunity to shift and turn to the deeper nature of life, whether it is through meditation, prayer, or contemplation, and invoke what I call an awakened awareness. We can move away from the narrow view of, "I have got to have it. I planned for it. It was mine." We can replace that with the view of, "What God, what Spirit, what Life, are you showing me now? You are showing me a bright open yellow door. A yellow door that is not only easy to open, I can walk right through to discover a landscape that I never knew existed before."

From the remnant of achievement awareness, I have only the information from today, and backward in time. But from the discovery of an awakened awareness, there is infinite information that I have yet to unpack and discern. The discovery of yellow doors in our lives is a discovery of a guided reality, opening up to a landscape that we may not have even known existed.

What happens at that junction between the stuck red door and the open yellow door? We have asked hundreds of people about this experience. They often speak of it in terms of synchronicity—that suddenly someone showed up in their path and pointed to the yellow door. "Suddenly, I found a book. Suddenly, I found a free flight or a very inexpensive flight to a part of the world I did not ever think I would visit. Suddenly, something changed."

As we look back on the moments in our lives when we lost the red door and gained the yellow door, most often people say there were messengers, there were helpers, there were healers. I call them "trail angels," who show up on the path. When we stand back even further and ask people to reflect on the yellow door, we hear them say, "I could feel God's presence in bringing the trail angel, I could feel the presence of life or spirit in guiding me to the yellow door."

Spiritual Growth and the Brain

After moments of hairpin curves, our paths lead to a form of recovery from loss, recovery from despair, and even trauma, through which we augment our awakened awareness. We have not merely recovered, we have not merely returned to baseline, but we are renewed and made bigger inside. A spiritual response to loss and trauma through prayer, meditation, and a deeper reflection and realization, leads us to know that we are loved and held. We are guided, and we are never alone.

The graph in **Figure 4** was drawn with over 3,000 people recovering from trauma. Everyone here met clinical criteria for post-traumatic stress disorder (PTSD). In this time of COVID, I invite you to think of how many people you know who walk with a sense of trauma. Yet what we discover is that as we move along the x-axis, we see more and more trauma.

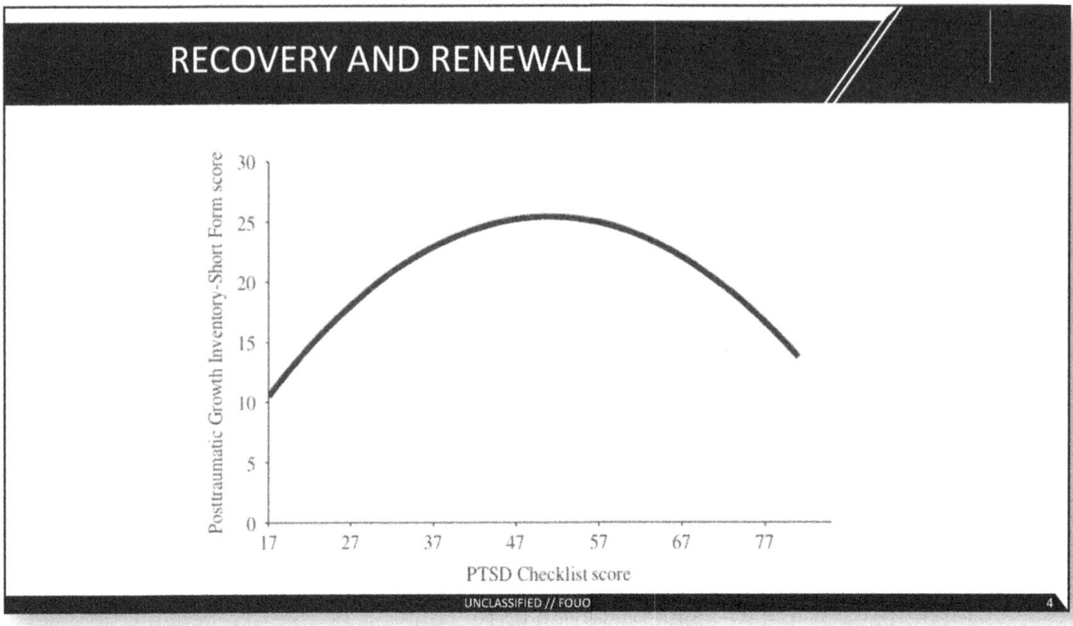

Figure 4

As we go up the y-axis, we see more and more growth, post-traumatic spiritual growth. So much so is this the case that 75% of people who met the criteria for PTSD at the same time reported meaningful growth.

Figure 5 outlines a realization of our awakened awareness and igniting of our spiritual core, so that on the other side we are in a deeper relationship of life. We see that we are not makers of our path, but we are discoverers of our journey. We are loved and held, we are guided, and we are never alone. We look at one another as trail angels, with love of neighbor, relational spirituality, and we draw closer to our Higher Power.

How is that? Consider your spiritual brain. Back to the big yellow circle: each one of us is born a spiritual being; each of us is equipped with a spiritual brain. However, whether or not we use our spiritual brain in moments of struggle and despair, when we are hitting our heads and fists on the red door, that is a choice. It is a choice to invoke the use of your spiritual brain through which you might then see the yellow door. We published this in 2014 in *JAMA Psy-*

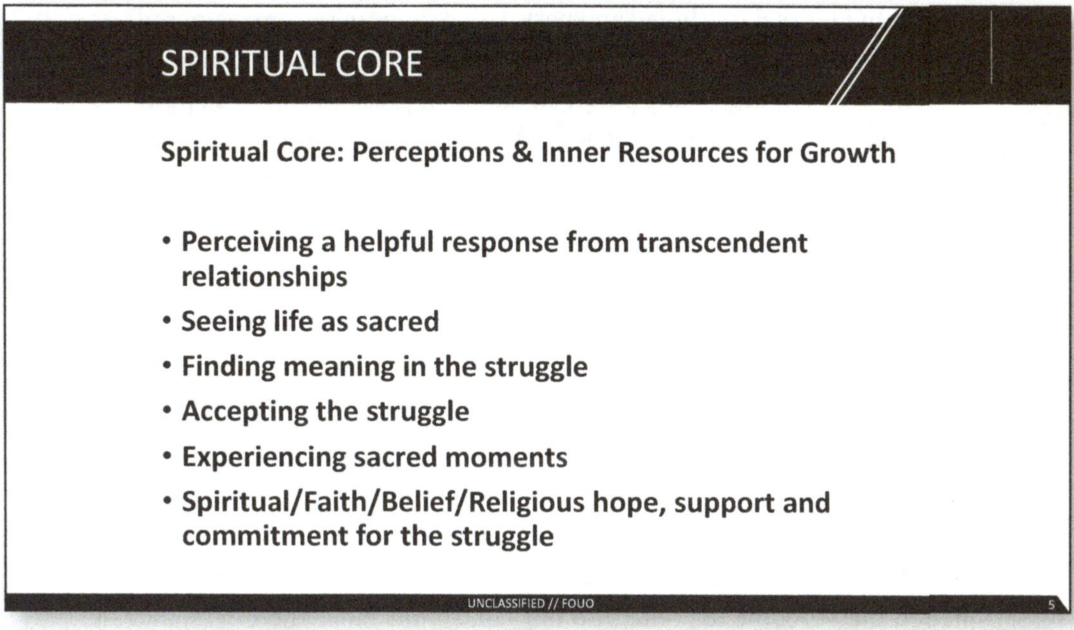

Figure 5

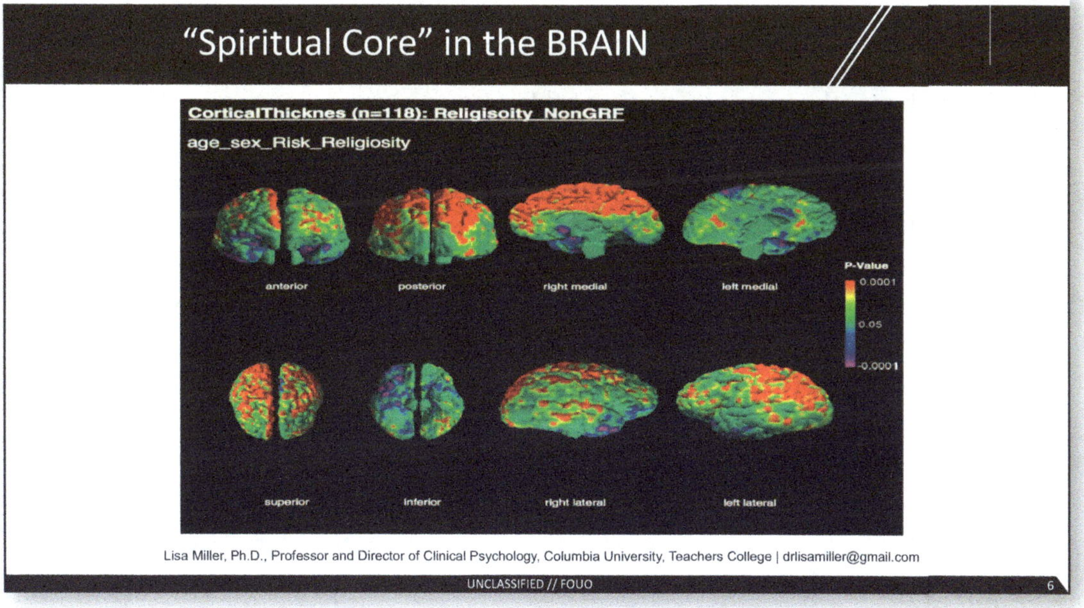

Figure 6

chiatry. In **Figure 6**, the broad and pervasive regions in red are regions of cortical thickness, regions like thick trees in a forest that are powerful and strong.

The cortex is processing power, and there is greater processing power in people with a sustained way of digging deep in times of suffering, people who in their darkest hour say, "What, God, do you show me now? What, Life, are you showing me now? What is the unitive reality? Where in the ocean am I being brought?" This is when we respond to suffering with an awakening and invoke our natural spiritual brain for awakened awareness. When we do, time and time again, it becomes the go-to place—it becomes the new normal. We invited these same people with red brains, a strong, awakened awareness, a spiritual way of seeing life's disappointments as opportunities, a spiritual way of understanding an impasse, as a call to shift and renew and go in another direction.

The Wavelength of All Life

The dialogue with life is a way of being, a living, dynamic dialogue. We invited these people to come into our lab, simply close their eyes, and sit down. What we found is that, time and time again, using awakened awareness, they gave a wavelength off the back of their head. This wavelength, a high-amplitude alpha of the spiritually engaged brain, is the wavelength that goes by another name in another field.

High-amplitude alpha is the same wavelength as Schumann resonance. It is the wavelength from the Earth's crust up one mile. It is the wavelength of nature. It is the wavelength of creation. Recovery from despair through an awakening of our naturally spiritual brain gives off the wavelength that is found throughout creation (**Figure 7**). The felt sense of being one with all life is mirrored in the oneness of this wavelength of energy.

Now I think that my grandchildren someday will laugh. They will laugh till they cry that their grandmother talked about wavelength, alpha, and Schumann resonance, when it is so abundantly apparent to them that there is one field of consciousness, one sacred field of life in and through all creation.

Today, however, we cross the bridge from radical materialism to post-materialism. Today, we move as a society, from thinking that only what we

Figure 7

can touch is real to using the multiple forms of awareness in our awakened awareness, to feel the unitive presence, the ocean sustaining all whitecap waves. It is a bridge to say that there is a common wavelength associated with knowing, feeling, and experiencing our unity with all life, and the wavelength is identical, found in all life.

It is a small step even for today, to say that the same wavelength perhaps reflects an ontology, a way of knowing and being, in a unitive reality. This is a unitive reality in which we are in relationship with water, with trees, with the Sun and air, with mountains, and yes, with little baby geese.

I invite you to share in this science and to share more deeply in the story of our finding of our precious son Isaiah in the book, *The Awakened Brain: The New Science of Spirituality and Our Quest for an Inspired Life.*

Thank you for the joy of sharing this work with you. Thank you for being on this journey to renew our global culture, so that we can live and connect as one emanation of the great Source of all life. ◊

About the Speaker

- Professor of Psychology and Education, Teachers College, Columbia University, New York
- Founder and Director of the Spirituality Mind Body Institute, Columbia University
- Founding Board Member, Academy for the Advancement of Postmaterialist Sciences
- Scholarly interests: Spirituality, Mental Health and Thriving: basic science and clinical science on spirituality, neural correlates and biological correlates of spirituality, spiritually integrated treatment and wellness interventions.
- Author of *The Spiritual Child: The New Science of Parenting for Health and Lifelong Thriving;* and *The Awakened Brain: The New Science of Spirituality and Our Quest for an Inspired Life;* editor of *The Oxford University Press Handbook of Psychology and Spirituality*
- PhD in Psychology

ICSG III

SESSION 1

Heralding a New Scientific Revolution

About Session 1

Heralding a New Scientific Revolution

Until recently, Western science has been based on a materialist philosophy, with the premise that we humans and the whole universe are simply a collection of particles and certain types of energy. In this view, consciousness is a byproduct of activities in the brain, and all entities exist separate from one another. However, it is unable to explain a wide range of anomalous or "extraordinary" phenomena, such as energy healing, channeling, and experiences during meditation. Therefore, there is a need for scientists to shift to a postmaterialist paradigm, in which consciousness plays a fundamental role, connecting all people and things. In order to identify the most effective technologies to address environmental problems, we decided to examine the applications arising from both frontier science, based on a postmaterialist perspective, and conventional science, based on materialism. In this session, the first speaker illuminates the features of a postmaterialist viewpoint, which is heralding a new scientific revolution. The second speaker presents the results of research into anomalous phenomena, demonstrating the need to adopt this new paradigm. ◊

About the Chair, Professor Menas Kafatos

- Fletcher Jones Endowed Chair Professor of Computational Physics, Chapman University, Orange, California, USA
- Director of the Center of Excellence in Earth Systems Modeling and Observations (CEESMO)
- Author or coauthor of books in many languages, including: *The Conscious Universe; The Non-local Universe* (with Robert Nadeau); *Principles of Integrative Science* (with Mihai Draganescu); *You Are the Universe* (with Deepak Chopra); and *Science, Reality & Everyday Life*
- PhD in Physics, Massachusetts Institute of Technology

Session 1 | Presentation One

Characteristics of the Postmaterialist Era

Professor Marjorie Woollacott
Institute of Neuroscience, University of Oregon, Eugene, USA

Two years ago, a wildfire in the Oregon State Forest came very close to my home in Eugene, and destroyed many homes and took many lives. It brought home for me the reality and immediacy of climate change, in its effect on my own life and that of my neighbors. This is reflected in some of the headlines regarding global warming you may have noticed in the last year from the US National Aeronautics and Space Administration (NASA), including the following:

- "2021 Tied for 6th Warmest Year in Continued Trend, NASA Analysis Shows" (NASA 2022a)

About the Speaker

- Professor Emeritus, Department of Human Physiology, and a member of the Institute of Neuroscience, University of Oregon, Eugene, Oregon
- President and Founding Board Member, Academy for the Advancement of Postmaterialist Sciences
- Research Director, International Association for Near-Death Studies (IANDS)
- Member of the Steering Committee of the Galileo Commission–Expanding the Scope of Science
- Author of *Infinite Awareness: The Awakening of a Scientific Mind* (2015)
- PhD in Cell and Molecular Biology; Research and teaching in neuroscience, child development, aging, rehabilitation, complementary and alternative medicine, and meditation

- "Sea Level to Rise up to a Foot by 2050, Interagency Report Finds" (NASA 2022b)
- "Reducing Emissions to Lessen Climate Change Would Yield Dramatic Health Benefits by 2030" (Candanosa 2021)

One of the first questions that comes to mind for me upon seeing these headlines is, "What can I do to help prevent climate change and thus maintain or improve the health of all of us on this Earth?" Today we will focus on this as we contemplate the possible characteristics of a postmaterialist era in which the nature of the environment would be optimal for the health of everyone. When I started to contemplate this topic, I asked myself these questions:

- How does a worldview (both individual and society) impact our relationship to the environment?
- What would be the characteristics of human consciousness, that is, the nature of our perception, behavior, and interactions, that would help bring about a postmaterialist era?
- And what does science say about how we might transform human consciousness in order to meet these expectations?

Let us first look at the question, "What is a worldview?" It can be defined as a set of beliefs about fundamental aspects of the nature of reality and our place in reality. A worldview creates the lens through which we view the world and how we perceive our place within it. Thus, your worldview influences all your perceiving, thinking, knowing, and doing. Most importantly, science tells us that we are typically unaware of our worldview, and worldviews are very hard to change (Koltko-Rivera 2004).

Characteristics of a Postmaterialist Era

The materialist view that is currently dominant in our culture holds that matter is the fundamental substance or reality of nature. Thus, all things, including the mind and consciousness, are secondary and products of material interactions. When the body dies, our consciousness is also destroyed. This leads to a sense of separation and competition for limited resources. It also contributes to a lack of concern for the well-being of future generations who must live on a healthy planet in order to thrive.

When we go beyond materialism to a postmaterialist worldview, the mind and consciousness are considered first-order realities, to which matter is subject and secondary. Thus consciousness, being primary, has the ability to affect and transform the material world. In a postmaterialist world, research is accepted as valid on topics such as nonlocal awareness, mystical experiences, and energy healing. The data also often opens us to new understandings of the nature of reality, interconnectedness, and healing.

What would be the characteristics of human perception, behavior, and interactions in a postmaterialist era? I would propose the following: the environment would be optimal for the health of the plants, animals, and humans who make the Earth their home. This would be brought forth through a new perception of unity and thus compassion for all beings on the planet, among human beings. This compassion would engender a sense of responsibility to care for our earthly home.

We might ask, "What exactly is this inner state of being, which would be the source of changes in the environment, and how would we experience it?" In fact, this experience is not uncommon today, experienced by those who meditate or are deeply immersed in the arts or athletics. Here are a few perceptions of this type of experience from different individuals:

Explorer Richard Byrd said of his sojourn in the Arctic, "I could feel no doubt of man's oneness with the universe. It was a feeling that transcended reason" (Mayer 2007). Patrick Vallençant, a champion French skier, commented, "At the beginning of any steep descent, there is a man and a slope of snow in unison...I become a part of this cosmic dimension" (Mayer 2007). A professor at a major university said the following after he began to meditate: "[a]fter the meditative awakening, I had an increased sensitivity, especially to nature–a sense of aliveness: I recall how I marveled as I looked at the world: everything was sharper, there was a greater luminosity, vibrant color. There was an increase in the feeling of love for my family and others. It was like a veil had been lifted off" (Woollacott and Lorimer 2022).

From these we see there are many paths to unity awareness, seeing humankind and nature as a seamless entity, through total immersion in nature and through stilling the mind. For example, in meditation, these people did not think their way to unity awareness. To do this, we must have the experience. When enough people in society have this experience, then there is a shift at the cultural level, in terms of how we view the environment, and the reverence we have for the natural world. We might conclude that this state may include an increased sensitivity and experience of oneness with nature, as well as a sense of connection and deep love for others. How do we shift our cultural awareness to a postmaterialist worldview?

One Inherent Source of a Materialist Worldview: The Filters Within the Brain

First, we need to understand one source of our materialist worldview: the filters within our brain obscure an expanded, nonlocal or unity awareness. They create the illusion that we are separate from others and have a transitory existence. Then we need to reduce the filters' activity, and, from this wider awareness, employ our "intuitive intelligence" to begin to solve the environmental challenges facing us (Mayer 2007). What are the filters in our brain that have helped create a materialist view of the world?

The first filters in our brain are the sensory receptors themselves that canalize the vibratory information that exists all around us. One example is the pigments in the visual receptors, which pick up only three ranges of wavelengths out of the whole vibratory spectrum—blue, green, and red (**Figure 1**). The same is true for all of the physical senses.

Research has also shown that our normal waking state is highly influenced by the Mind Wandering Network or Default Mode Network (DMN) of the brain, showing the high levels of activity when we are normally thinking (**Figure 2**). It creates the narrative or stories we tell about ourselves and others throughout our day. The DMN is considered by many to be the

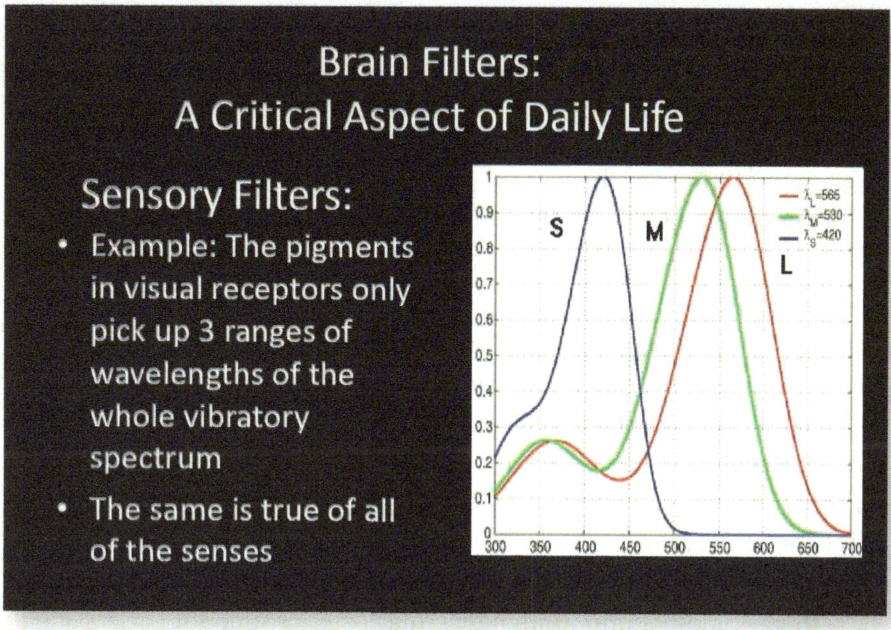

Source: https://commons.wikimedia.org/wiki/File:ColorVision.jpg **Figure 1**

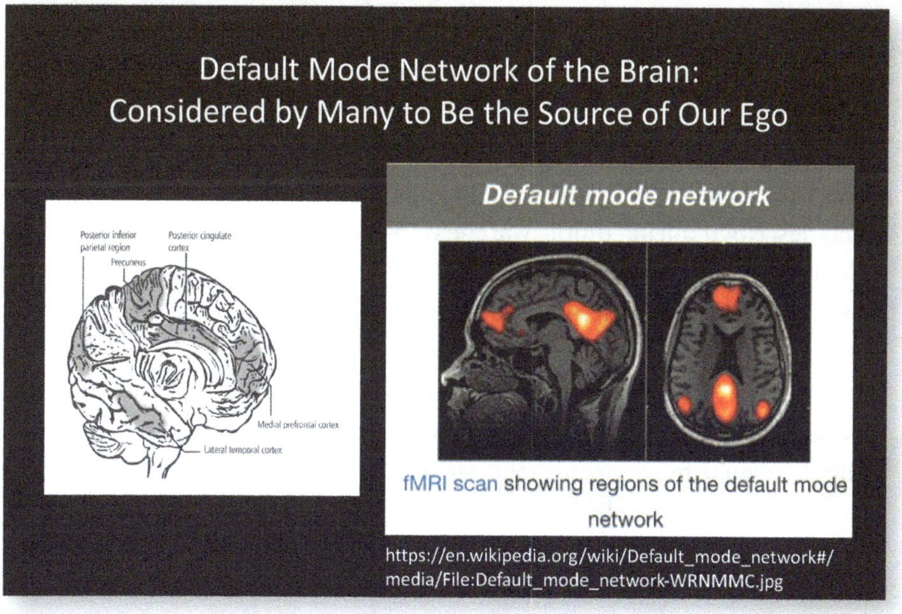

Left image source: Author (https://marjoriewoollacott.com/wp-content/uploads/2021/07/Woollacott-Shumway-Cook-NEURAL-CORRELATES.pdf; Woollacott and Shumway-Cook 2020; adapted from Ricard et al. 2014) **Figure 2**
Right source: https://en.wikipedia.org/wiki/Default_mode_network#/ media/File:Default_mode_network-WRNMMC.jpg

source of the ego—the notion of self and our identification with the thoughts, emotions, and roles we play in the world. The DMN's main influence is to filter out a broader perceptual awareness, so that what we experience ends up being a small trickle of the information that is available from the universe. The more we are involved in our egoic narrative, the less we see of the real nature of the world. During day-to-day living it is very active.

In addition to the narrative filtering of the DMN, the left hemisphere networks involve analysis, language processing networks, and the like in the brain and are important conceptual filters. They reduce our awareness of holistic perceptions of the right hemisphere networks such as creativity, imagination, and holistic thinking. In fact, left-hemisphere language centers have been described as a filter of perception, focusing attention on particular aspects of the world. McGilchrist (2009) proposes that in our mechanistic age, the left brain is dominant, and thus more holistic perceptions of reality are diminished.

There is another important filter—the posterior superior parietal lobule of the brain—that limits our sense of space and time. Dr. Andrew Newberg has shown that in experienced meditators, who said they experienced an expansion of consciousness and unity awareness during deep meditation, there is a reduction in activity in this part of the brain. Newberg hypothesizes that a reduction of the sensory and cognitive inputs into this area during meditation is associated with the sense of going beyond space and time that is so often described during unity awareness (Newberg and D'Aquili 2001).

The last filters are the thalamus and the thalamo-cortical loop. Our Executive Attention system in the prefrontal cortex would get overwhelmed with information if the thalamus passed along all sensory inputs. However, when it is more important to pay attention to one sense, such as hearing versus sight, there is inhibition of the visual part of the thalamus, leaving the auditory signal more prominent. In this case, the prefrontal cortex then instructs the basal ganglia to activate the thalamic reticular nucleus to inhibit the visual part of the thalamus. This then suppresses the flow of visual information and leaves the auditory signal more prominent. Another way that the thalamus takes this filtering even further is by reverberating thalamo-cortical internal circuitry that actually competes with sensory inputs for attentional focus (Llinas and Paré 1996). This is an internal loop from the cortex to the thalamus and back, which can dominate all our perceptions with previous memories and our verbal stories. This means that, under normal conditions, brain inputs from our current thoughts can be more powerful than inputs from the world around us in creating our perception of reality.

In summary, the brain's filters are critical to our ability to function in everyday life; without the filters we would be overwhelmed. Our brains are hardwired to predispose us to an experience of separation; thus, we are not hardwired for unity awareness.

Reducing the Effect of the Brain's Filters

Are there ways to reduce the filters so that we can experience our expanded awareness and connectedness with others? We can do this by stilling the DMN and other filters in the brain through a variety of ways, including meditation. Let us examine a case study that explores perceptual changes during a meditation practice. Roger Walsh (1983), a professor of Psychiatry at University of California, Irvine, writes in a journal article:

> "[After deep meditation,] the experience feels like having a faint but discernible veil removed from my eyes, and that the veil is made up of hundreds of subtle thoughts and feelings. Each one of these thoughts and feelings seems to act as a competing stimulus or 'noise' that thus reduces sensitivity to any one object. Thus, after meditation, any specific stimulus appears stronger and clearer, presumably because the signal:noise ratio is increased."

This portrayal of post-meditation perception as removing a "discernible veil" consisting of layers of mental and emotional activity that compete for one's attention is a compelling metaphor for reduction in the function of the DMN.

Research by Judson Brewer and his colleagues from the Department of

Psychiatry, Yale University School of Medicine, gives quantitative evidence as well. Using brain imaging (fMRI), the study showed significant reductions in two primary nodes of the DMN of experienced meditators vs novices during mystical states of meditation. They found that the main nodes of the network are relatively deactivated in meditators compared to novices during meditation. The medial prefrontal cortex and the posterior cingulate cortex exhibited reductions in activity in the meditators versus the controls in both of those areas of the brain. Further studies showed reductions additionally in widespread areas of the brain (Brewer et al. 2011; Garrison et al. 2015). This shows us that the brain truly becomes quite still in meditation as the narrative part of the brain is silenced. I propose that this stilling of our narrative mind actually allows subtler perceptions to come through our awareness that are normally filtered out, that is, the perceptions that mystical experiences are made of.

Awakening to the Experience of a Postmaterialist Worldview: The Effects of Meditation

Research studies have explored many tools to help accelerate the shift in our individual and cultural awareness to one of a postmaterialist worldview. One is the practice of meditation. Many studies have shown that the practice of meditation increases our compassion and empathy for ourselves, others, and the planet. For example, Paul Condon and colleagues (2013) explored whether meditation training would make you more likely to come to the aid of a person in pain even when others were ignoring the person. Condon and colleagues asked participants to come to the lab for what they were told was a test of cognition upon their completion of an eight-week meditation training. The experiment was set up so that two people were already in two of the chairs in the waiting room, and the participant takes the last chair that is left. Then a person comes on to the floor in crutches, in obvious pain, while the first two persons ignore him. The question is whether the participant will give up the open chair for the person in pain. Results showed that meditation caused a significant increase in compassion or empathy, compared to the control group, with 50% of meditators and only 15% of non-meditators acting to help. Meditation exhibited a tremendous effect. This clearly shows that the practice of meditation results in strengthening the capacity to experience and express compassion.

Scientists, such as Dr. Richard Davidson at the University of Wisconsin, have also shown that this increase in compassion is associated with changes in brain structures, including increased activity in the dorsolateral prefrontal cortex and increased activity in the anterior cingulate cortex, associated with positive feelings and decreased activity in the amygdala, which is associated with distress. Davidson made an important point about this type of study. He said it shows that, rather than becoming depressed by suffering, people who are trained in meditation develop a strong disposition to alleviate suffering and to wish others to be happy (Davidson and Begley 2012).

Can meditation training also improve health care itself? Let us now look at the applications of this research to the health-care setting. One study by Claudia Orellana-Rios and her colleagues (2018) in Germany trained a palliative care team in meditation, which was conducted at their workplace. The quantitative data showed that the staff met or significantly exceeded their goals, including experiencing increased peace in stressful situations. For example, one nurse said,

> "I stopped just for a short moment in very chaotic or critical situations and got myself out of that sort of vortex we can very quickly get in to here, where you start thinking all this is so horrible and awful and too much and when you take a step back, you notice it's not necessary to take it upon you. And then the situation sort of widens and changes. I did that all the time."

Here is also one study on applications of mindfulness to early childhood education, from Richard Davidson's lab at the University of Wisconsin (Flook et al. 2015). They conducted a 12-week mindfulness-based Kindness Curriculum that was taught in a public preschool setting. The training focused on improving executive function, self-regulation, and prosocial behavior. They found significant increases in learning, health, and social/emotional abilities in the mindfulness/kindness curriculum group versus a control group. One example was that the mindfulness group acted less selfishly, that is, more selflessly, than the control group over time, sharing more of their resources. This suggests there may be benefits in incorporating mindfulness-based training into an early education curriculum.

Finally, there are also applications of mindfulness in the government/legislative setting.

In the United Kingdom, legislators formed the Mindfulness All-Party Parliamentary Group (MAPP) to implement mindfulness into health, education, workplace, and the criminal justice system (Mindfulness Strategies).

The MAPP group says that these programs have assisted government workers in developing healthier, more compassionate responses to their own experience, as well as to events in their lives and the people who are in them. In the United States, Representative Tim Ryan has been an advocate for mindfulness in the government. We have shown evidence that mindfulness is effective in shifting our worldview to one of unity, compassion, and connectedness, and can be applied in a variety of settings.

Barriers to Overcome

Using this evidence as a foundation, we can create a new narrative about humanity and how we relate to the world. We can go to the source (changing our awareness and understanding) to solve environmental problems and not just the periphery. What is the key to changing our worldview to a post-materialist perspective? I believe it is having a direct experience of expanded consciousness. How do we facilitate this?

First, we need to remember that worldviews are largely unconscious, and we need to become more aware of our worldviews and their effects on our behavior. One of the main barriers to overcome in making this shift is our current very solid materialist worldview we were raised in, such as in our western science-oriented culture. Here is what one scientist, Dr. Harold Puthoff, said about this when he began doing research on the existence of nonlocal aspects of consciousness and found that his subjects in his experiments at the Stanford Research Institute could actually see things clearly in their own minds that were thousands of miles away, as if they were present:

> "The evidence we had was rock hard. I saw that. But I also saw that it didn't eradicate my doubt. That made me see my doubts weren't the problem. On the contrary, the problem lay with my beliefs. I was having terrible trouble giving up my beliefs about how the world worked, even in the face of evidence that said my beliefs were wrong. It really brought home what any good scientist knows: doubts aren't what interfere with good science... It's easier to think there must be something wrong with your experiments. And that's ironic because it looks like humility, but it isn't. It's privileging your own personal beliefs above the evidence. That's not science, it's hubris" (Mayer 2007).

Mayer (2007) summarizes this and other barriers of many scientists to embracing a postmaterialist era. They include: 1) being trained in a solid materialist belief system, which is the case for most scientists; 2) the fear of letting go of this lifelong societally dominant belief; 3) the fear that these experiences are illusions; 4) the fear of being considered mentally unbalanced by family, friends, colleagues; 5) the fear of ruining one's career; and 6) resistance and ridicule by those who have never had this type of experience.

How can we encourage these experiences and remove the barriers to embracing a postmaterialist era? The research I summarized earlier suggests that within the educational system, we can teach children, starting in early education, to quiet the mind (Flook et al. 2015). We can also train health-care workers in mindfulness meditation (Orellana-Rios et al. 2018), as well as offer training in mindfulness to a country's governing bodies. In addition, of course, we can create environments that induce relaxation and stillness, and encourage workers to take time during the workday to spend time there in stillness. We can also earmark funding for further studies on these important effects of meditation, on increasing a sense of interconnectedness, and a disposition to alleviate suffering in others.

In summary, the belief that changing the worldviews of both individuals and society is KEY to changing our attitudes, beliefs, and actions towards the environment. We cannot "save" the environment of this planet without changing the underlying worldview from a materialist worldview related to humankind's right to exploit the environment to a postmaterialist worldview of humankind's need to protect the environment. These shifts in worldview may originally happen one person at a time, as we learn the value of quieting the mind and expanding our awareness—but at some point, I believe that our culture will reach a tipping point in which we truly understand, value, and put into practice this new attitude toward ourselves and the planet. It is already beginning to happen, not only in our individual lives but also in health care, in early childhood education, and even in governmental organizations. ◊

References

Brewer, J.A., P.D. Worhunsky, J.R. Gray, Y.Y. Tang, J. Weber, and H. Kober. 2011. "Meditation experience is associated with differences in default mode network activity and connectivity." *Proceedings of the National Academy of Sciences USA* 108 (50): 20254–20259. https://doi.org/10.1073/pnas.1112029108

Broadbent, D.E. 1958. *Perception and Communication.* Oxford, UK: Pergamon Press.

Candanosa, R.M. 2021. "Reducing Emissions to Lessen Climate Change Would Yield Dramatic Health Benefits by 2030." NASA. Accessed Feb 18, 2022. https://climate.nasa.gov/news/3134/reducing-emissions-to-lessen-climate-change-would-yield-dramatic-health-benefits-by-2030/

Condon, P., G. Desbordes, W.B. Miller, and D. DeSteno. 2013. "Meditation Increases Compassionate Responses to Suffering." *Psychological Science* 24 (10): 2125–2127. https://doi.org/10.1177%2F0956797613485603

Davidson, R.J., and S. Begley. 2012. *The Emotional Life of Your Brain: How Its Unique Patterns Affect the Way You Think, Feel, and Live—and How You Can Change Them.* New York City, NY: Avery Publishing.

Driver, J. 2001. "A selective review of selective attention research from the past century." *British Journal of Psychology* 92 Part 1: 53–78.

Flook, L., S.B. Goldberg, L. Pinger, and R.J. Davidson. 2015. "Promoting prosocial behavior and self-regulatory skills in preschool children through a mindfulness-based kindness curriculum." *Developmental Psychology* 51 (1): 44–51. https://psycnet.apa.org/doi/10.1037/a0038256

Garrison, K.A., T.A. Zeffiro, D. Scheinost, R.T. Constable, and J.A. Brewer. 2015. "Meditation leads to reduced default mode network activity beyond an active task." *Cognitive, Affective and Behavioral Neuroscience* 15: 712–720. https://doi.org/10.3758/s13415-015-0358-3

Huxley, A. 1954. *The Doors of Perception.* New York City, NY: Harper & Brothers.

Koltko-Rivera, M.E. 2004. "The Psychology of Worldviews." *Review of General Psychology* 8 (1): 3–58. https://doi.org/10.1037%2F1089-2680.8.1.3

Llinás, R.R., and D. Paré. 1996. "The Brain as a Closed System Modulated by the Senses." In *The Mind–Brain Continuum: Sensory Processes,* edited by R.R. Llinás and P.S. Churchland, 1–18. Cambridge, MA: The MIT Press.

Mayer, E.L. 2007. *Extraordinary Knowing: Science, Skepticism and the Inexplicable Powers of the Human Mind.* New York, NY: Bantam Books.

McGilchrist, I. 2009. *The Master and his Emissary: The Divided Brain and the Making of the Western World.* New Haven, CT: Yale University Press.

Mindfulness Strategies. Accessed Feb 18, 2022. https://www.mindfulnessstrategies.com/

NASA. 2022a. "2021 Tied for 6th Warmest Year in Continued Trend, NASA Analysis Shows." Accessed on July 7, 2022. https://climate.nasa.gov/news/3140/2021-tied-for-6th-warmest-year-in-continued-trend-nasa-analysis-shows/

———. 2022b. "Sea Level to Rise up to a Foot by 2050, Interagency Report Finds." Accessed on July 7, 2022. https://www.nasa.gov/feature/jpl/sea-level-to-rise-up-to-a-foot-by-2050-interagency-report-finds

Newberg, A., E. D'Aquili, and V. Rause. 2001. *Why God Won't Go Away: Brain Science and the Biology of Belief.* New York City, NY: Ballantine Books.

Orellana-Rios, C.L., L. Radbruch, M. Kern, Y.U. Regel, A. Anton, S. Sinclair, and S. Schmidt. 2018. "Mindfulness and compassion-oriented practices at work reduce distress and enhance self-care of palliative care teams: a mixed-method evaluation of an "on the job" program." *BMC Palliative Care* 17 (3): 3–18. https://doi.org/10.1186/s12904-017-0219-7

Walsh, R. 1983. "Meditation Practice and Research." *Journal of Humanistic Psychology* 23 (1): 18–50. https://journals.sagepub.com/doi/10.1177/0022167883231004

Woollacott, M., and A. Shumway-Cook. 2020. "The Mystical Experience and Its Neural Correlates." *Journal of Near-Death Studies* 38 (1): 3–25. Accessed on July 7, 2022. https://marjoriewoollacott.com/wp-content/uploads/2021/07/Woollacott-Shumway-Cook-NEURAL-CORRELATES.pdf

———, and D. Lorimer. 2022. *Spiritual Awakenings. Scientists and Academics Describe Their Experiences.* Battle Ground, WA: AAPS Press.

Session 1 | First Commentary
on the Presentation by Professor Marjorie Woollacott

Characteristics of the Postmaterialist Era

Professor Janice Miner Holden
Department of Counseling & Higher Education,
University of North Texas, Denton, USA

It is my great pleasure to comment on Dr. Woollacott's paper and speech, "Characteristics of the Postmaterialist Era."

Her main points were:

- to describe the difference between a materialist and a postmaterialist worldview;
- to hypothesize that widespread human adoption of a postmaterialist worldview would result in a shift from a basic attitude of exploitation to one of protection of Earth's environment, which would, in turn, herald a new era of compassion for and well-being of Earth and its inhabitants;

About the Speaker

- Professor Emerita; Former Chair, Department of Counseling & Higher Education, University of North Texas, Denton, USA
- President, International Association for Near-Death Studies (IANDS)
- Editor-in-chief, *Journal of Near-Death Studies* (Since 2008)
- Texas Licensed Professional Counselor-Supervisor and American Center for the Integration of Spiritually Transformative Experiences (ACISTE) Certified Mental Health Professional
- EdD, LPC-S, ACMHP

- to explain various factors that impede transition from the currently dominant materialist worldview to a postmaterialist perspective, including inherent filter functions in the human brain;
- to present evidence that the practice of meditation reduces the filter functions of the brain and thus promotes a shift toward dispositions that characterize a postmaterialist worldview;
- to identify barriers to shifting to a postmaterialist perspective; and
- to suggest that the most effective way to promote such a shift is for individuals to have a direct experience of expanded consciousness, in particular through the practice of meditation.

The overall theme of my commentary is, "Yes, and...." That is, I agree with everything Dr. Woollacott presented, and I would like to address further the question of how to facilitate a widespread shift from a materialist to a postmaterialist worldview (in other words, "the shift,") by adding one point to what Dr. Woollacott said and expanding on another point that she made.

I believe that people undergo the shift through one or both of two processes: one logical, and the other experiential. My first, additional point relates to the logical route. Reasoning represents the primary source of knowing for most people over the age of about 7 years (Wilber 2000). The data with the most persuasive potential are those that derive from empirical investigation. Extensive empirical research now points to a postmaterialist perspective that consciousness functions beyond the confines of the brain.

In the past few years, several organizations have undertaken as their mission to present these data to the public. Examples include the activities and publications of the Scientific and Medical Network (https://scientificandmedical.net/; https://galileocommission.org/; https://scientificandmedical.net/events/dr-mario-beauregard-expanding-reality-the-emergence-of-post materialist-science/) and the Academy for the Advancement of Postmaterialist Sciences (https://www.aapsglobal.com/). The International Association for Near-Death Studies, for which I serve currently as president, recently held a two-day symposium in which leading researchers on a variety of phenomena presented empirical data indicating the ability of the human mind to function independent of the brain. These phenomena included psychic experiences; veridical (verified) perception from near-death experiences, shared death experiences, and after-death communication (ADC); terminal/paradoxical lucidity; and evidential mediumship (https://symposium.iands.org/).

As these and other organizations, along with popular journalists, continue to bring this information to the general public, members of the public willing to consider such evidence may be persuaded by it. I have seen this process in action in a graduate course I developed and have taught for over two decades; students are usually completely unaware of the extent to which transpersonal experiences—those that transcend the usual personal limits of space, time, identity, and/or influence—have been researched, and once they encounter this research, they almost unanimously manifest the shift (Walker 2022). I believe that professional organizations, educational institutions, and journalists have the power to advance this shift on a societal scale by persistently presenting it to the public.

My second, expansive point relates to the second route to the shift: by actually having transpersonal experiences. This is the route to which Dr. Woollacott alluded when she recommended meditation as an avenue to worldview transformation. Transpersonal experiences themselves do not derive from logic; rather, they involve a holistic transrational process (Wilber 2000). This process is phenomenologically different from the process of reasoning. When I experience precognition, ADC, spontaneous mediumship, and healing following a past-life experience, these experiences occur as holistic knowings that not only do not arise from sensory or logical sources but, in many cases, defy logic—yet hold undeniable personal validity. Such experiences have the power to facilitate the shift.

Empirical evidence for this process comes, for example, from Michele Knight's (2011) qualitative study of 21 bereaved adults in Sydney, Australia, who experienced spontaneous ADC with their deceased loved one. She found that, following ADC, those who had previously believed in the survival of consciousness after death were reinforced in that belief, whereas those who had not previously so believed were now convinced of it.

In addition to traditional meditation as a widely available avenue to transpersonal experience, I would like to suggest a related possibility. As Dr. Woollacott indicated, a significant difference in brainwave activity has been observed between novice and experienced meditators—with the latter reporting more mystical experience. The evolution from novice to experienced meditator can involve a relatively inefficient trial-and-error process that often takes years.

As an alternative, neurofeedback technologies can facilitate this consciousness evolution (see, for example, Stockly and Wildman 2021). These processes involve measurement of brain activity via electrodes placed in a person's head area and connected to a machine that detects when brain activity is changing in the direction of reduced activity in the Default Mode Network—the condition associated with mystical experience—and provides feedback to the person, usually in the form of some audio stimulus such as a soft tone, when the person's brain is "going in the right direction." With such feedback, a person can more efficiently learn meditation—and therefore be more likely to potentiate the shift. Imagine a world in which such technology would be widely available—and perhaps, because of the general health benefits of meditation, would be government subsidized!

In summary, it is my hope that a postmaterialist era will be characterized by widespread availability of both empirical information about the nature of consciousness to appeal to people's logical function, as well as of technologies that facilitate expanded consciousness to appeal to people's experiential function. These two functions, working in concert, will hopefully advance a society in which a postmaterialist perspective is the prevailing worldview. ◊

References

Knight, M. 2011. “Ways of being: The alchemy of bereavement and communiqué.” PhD dissertation. University of Sydney. Accessed July 7, 2022. https://ses.library.usyd.edu.au/handle/2123/13764

Stockly, K., and W. Wildman. 2021. “Mystical experiences on demand? Welcome to spirit tech.” BIG THINK. Accessed on March 22, 2022. https://bigthink.com/neuropsych/spiritual-mystical-experiences-brain-tech/

Walker, U.N. 2022. "Transpersonal in counselor education: A phenomenological inquiry." *Dissertations, Theses, and Masters Projects.* Williamsburg, VA: William & Mary. https://dx.doi.org/10.25774/w4-wwar-7v26

Wilber, K. 2000. *Integral Psychology: Consciousness, Spirit, Psychology, Therapy.* Boulder, CO: Shambhala Publications.

Session 1 | Second Commentary
on the Presentation by Professor Marjorie Woollacott

Characteristics of the Postmaterialist Era

Dr. Diane Hennacy Powell

Neuropsychiatrist, Independent Researcher, Treasurer, Parapsychological Association, Medford, Oregon, USA

Professor Woollacott began by telling us about the importance of our worldview. Although it informs our attitudes toward all aspects of life, she notes that it is hard to change. Why? It is often founded on unchallenged beliefs that are hard to articulate because they are so deeply rooted. When opposing views are presented to us, they are often met with hostility because acceptance can create cognitive dissonance, which is uncomfortable. Although the obstacles to change that she outlined are still operative, we are currently witnessing an unprecedented shift in worldviews. Advances in technology are driving this in part because they provide

About the Speaker

- Neuropsychiatrist; Independent Researcher; Treasurer, Parapsychological Association, Oregon, USA
- Integrative medical doctor specializing in Alzheimer's, Autism, Major Depression, and PTSD
- Research in neuroscience of extraordinary states of consciousness and anomalous experiences
- Author of the book, *The ESP Enigma: The Scientific Case for Psychic Phenomena* (2008), and contributor to *Seriously Strange: Thinking Anew about Psychical Differences* (2012) and other books

widespread access to countless other cultures and their perspectives, but this transition is also arising because the technology itself has been revolutionary. The forces behind this evolution have their downside, but where they are taking us is generally a welcome trend, and our survival as a planet and a species might depend upon it.

The postmaterialist worldview presented by Professor Woollacott is gaining traction in some academic circles largely because there is scientific evidence for both its validity and benefits, but the full fruits of adopting it are yet to be realized. If belief in our interconnectedness became the dominant worldview, it could bring about greater cooperation among us as global inhabitants. In other words, we might act differently if we realized that we cannot harm another without harming ourselves. Similarly, if we believed in reincarnation and karma, even the most selfish and materialistic people would feel more compelled to care about their actions and the future of the planet.

Worldviews are the result of early programming by our parents, teachers, and the media, with the latter playing an increasing role. Early programming has the most profound impact, but as we get older and our consciousness evolves, we can choose to either continue along the path of our youth or branch out to various degrees to express our individuality. This evolution has been happening at a faster pace over the past century because the messages in our media and schools have changed radically. As a result, the split in worldviews across generations has widened, and expectations are shifting.

As we go through our daily affairs, we need a narrative to make sense of our lives. We unconsciously fit events into our narrative, which could be ego-building, ego-destroying, or a combination of the two. It all depends upon the individual and their programming. Our brains often function like echo-chambers for the things we have been told to believe, which is why these repetitive thoughts and their origins are called programming. Recognizing this is essential to moving forward. Psychiatrists and neurosurgeons can tinker with the hardware (in other words, the brain) as much as they want, but when it comes to the causes of personal suffering and societal dysfunction, the software (programming) is often the real issue.

If you want to know what and how someone from a different generation, socioeconomic group, or other subculture thinks, read their books, listen to their favorite programs, and ask them relevant questions. Do they prefer fiction because of the emotional safety net of knowing that the world they are exploring is just someone's fantasy, or do they prefer nonfiction because they need to believe it is true to learn from it? Which narratives do they relate to because they identify with a character, role, or plot theme? Is that identification just their wishful imagination, or does it speak to their core sense of identity?

Prof. Woollacott's talk helps elucidate the role of the Default Mode Network (DMN) in our illusion of separation. It also describes how meditation can help individuals temporarily suspend or turn it off, which is why meditation is liberating. For example, if circuitous thoughts arise during meditation, they are allowed to float away, instead of continuing on in an unproductive inner dialogue. Initially meditation provides a brief respite from the noise, but over time the practice can literally rewire our circuitry. In other words, unlike our personal computers, the hardware of our brains can be modified by the software we run on them. In a truly post-materialist world, we will not just turn off the limiting belief system, but we will recognize that everyone creates their own reality, not as a cliché but as a scientific fact.

Prof. Woollacott presented evidence that meditation can help people find an inner peace and sense of oneness or unity. Once people have found this inner peace, they can become ambassadors for peace simply by embodying the characteristics of human consciousness that result from meditation. When we become hubs from which love is transmitted, we not only get to feel its full impact, we draw others to us. This is easy to understand. When we are compassionate, happy, and grateful, we treat everything and each other as blessings—with a love so infectious that everything else pales in comparison.

What about those for whom love is an elusive concept? It must be experienced. Perhaps this is because love has a specific frequency that must literally be felt to be accepted as real. Many of us have been so traumatized we need to learn how to trust love and its source, which is why it needs to begin within. After we experience self-love, we are better able to send love. Despite the obvious benefits of love, another obstacle is that many of us fear turning on the spigot of love out of the fear of loss. The answer to that conundrum is that we must learn to love and lose. That requires a surrender to forces beyond our control, which is why we might resist. However, once we, individually and collectively, realize our immortality, it will become apparent that our perceived loss is only in the material realm. It is then that we will be eternally free. ◇

Session 1 | Presentation Two

Anomalous Phenomena that Require Shifts in the Scientific Paradigm

Dr. Helané Wahbeh
Institute of Noetic Sciences, Petaluma, California, USA

In order to understand anomalous phenomena that require shifts in the scientific paradigm, we must first start with some definitions and descriptions of precisely what these anomalous phenomena are, how common they are, and the evidence for their validity. Then we can discuss the current scientific paradigm and why these phenomena require a shift in the paradigm.

Let us begin with "anomalous phenomena." Anomalous means "deviating from what is standard, normal, or expected." Scholars and researchers have studied these phenomena for over 150 years in parapsychology. Anom-

About the Speaker

- Director of Research, Institute of Noetic Sciences, Petaluma, California
- Adjunct Assistant Professor, Department of Neurology, Oregon Health & Science University, Portland, Oregon
- She has published on and spoken internationally about her studies on complementary and alternative medicine, mind-body medicine, extended human capacities, stress, post-traumatic stress disorder, and their relationships to physiology, health, and healing
- Author of *The Science of Channeling: Why You Should Trust Your Intuition and Embrace the Force That Connects Us All*; coauthor of *Free To Be Me: A Journey of Transformation through Generational Healing*
- Clinically trained as a naturopathic physician (ND) and research trained with a Master of Clinical Research (MCR)

alous phenomena are varied and include a range of experiences such as extrasensory perception, psychokinesis, and mediumship.

Extrasensory perception (ESP) refers to perceptions that are not mediated by our traditional five senses. Some people also use the term psychic to describe these experiences. Other terms commonly applied are telepathy, or mind-to-mind communication, and precognition, or perceiving the future before it happens. In addition, there is a whole range of words with the prefix "clair-" (meaning "clear, bright, distinct"), such as in clairvoyance (literally "clear seeing"), clairaudience (literally "clear hearing"), or clairsentience (literally "clear sensing"). These "clair-" terms attempt to describe a specific experience of perception beyond the traditional five senses. They encompass a sensory experience of accessing information that we would not usually know.

Another aspect of anomalous phenomena is expressive and is called psychokinesis (PK), or "mind over matter." There are three commonly described types of psychokinesis: macro, micro, and bio. Macro-PK is the mind's influence on objects that we can see with our eyes (Braude 2015). Micro-PK is the mind's influence on probabilistic systems that require statistical analyses to observe. Bio-PK refers to mind over living matter, such as human bodies, cells, animals, or plants.

Other anomalous phenomena include mediumship, where a person believes they are communicating with a deceased person or non-physical being. For example, the medium receives messages from the deceased person and then communicates them to the sitter. In trance channeling, the channeler believes the deceased person or non-physical being uses their body to communicate directly.

The many terms used to describe these "anomalous" experiences can sometimes create confusion. Michael Thalbourne's *A Glossary of Terms Used in Parapsychology* is a 130-page text of the various parapsychological terms. In the preface, the author notes, "[You] may have as much difficulty in reading the literature of the subject as if it were written in a foreign language" (Thalbourne 2003, xiii).

A broad term encompassing all these anomalous phenomena is chan-

12 Noetic Signature™ Characteristics

1. Inner Knowing
2. Embodied Sensations
3. Physical Sensations from Other People
4. Inner knowing through touch
5. Inner Voice
6. Apparent Non-Physical Being Communication
7. Knowing the Future
8. Knowing Other's Minds
9. Knowing Through Dreams
10. Affecting Your Personal World
11. Visualization/Manifestation
12. Healing

IONS

© Institute of Noetic Sciences

Figure 1

Credit for all figures: Courtesy of Dr. Helané Wahbeh and the Institute of Noetic Sciences

neling. I define channeling as the process of revealing information and energy not limited by our conventional notions of space and time and can appear receptive or expressive. As I alluded to above, channeling is experienced on a spectrum. On one side of the spectrum are the widespread experiences of intuition and gut hunches that are very common. On the other side are rare experiences, such as mediumship and trance channeling. I propose that accessing this information and energy from beyond our conventional notions of time and space is an innate capacity that all humans have. However, how the capacities express themselves is unique to them and what I call the Noetic Signature (Wahbeh et al. 2022; Wahbeh, Fry, and Speirn 2022) (**Figure 1**).

Now that we have reviewed some of the definitions and descriptions, we can summarize the prevalence studies on these anomalous phenomena (**Figure 2**). Researchers have been formally studying how common channeling is for about 40 years (Bourguignon 1976; Castro et al. 2014; Cohn 1994;

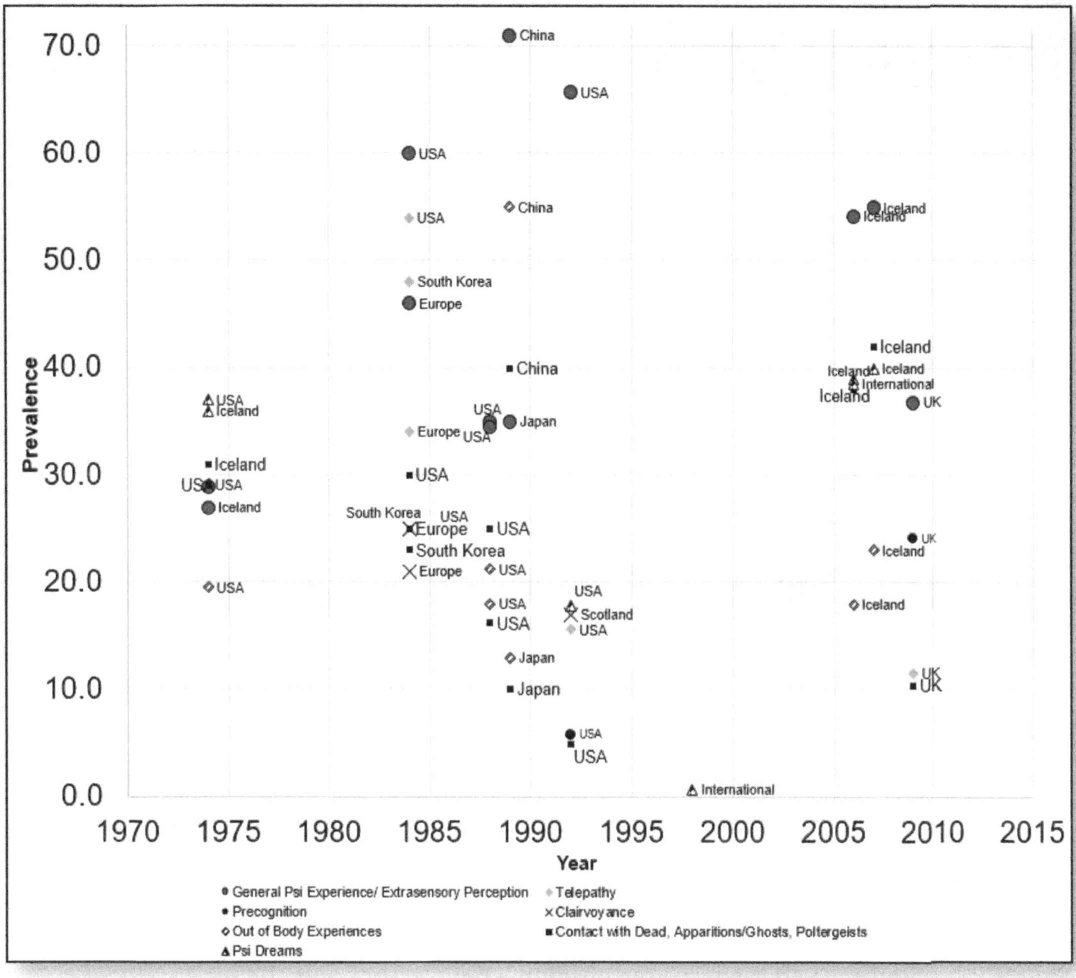

Figure 2

Greeley 1987; Haraldsson 1985, 2011; Haraldsson and Houtkooper 1991; McClenon 1993; Palmer 1979; Ross and Joshi 1992), with prevalence rates ranging from as low as 10% in Scottish citizens (Cohn 1994) to 97% in enthusiasts in the United States (Wahbeh et al. 2018). One study surveyed 1,120 meditators (**Figure 3**) about these phenomena and found that most of them had experienced clairvoyance and telepathy, alterations in time and space,

Figure 3

and communications with non-physical entities (Vieten et al. 2018). We also surveyed the general public, scientists and engineers, and channeling enthusiasts in the United States (Wahbeh et al. 2018) and found that over 90% had experienced at least 1 of 25 anomalous experiences (**Figure 4**). With prevalence rates being well-above 10% of most populations surveyed, we clearly witness that these phenomena are common. I would argue that these phenomena should not be called anomalous due to their widespread prevalence.

So these experiences are common—but are they actually real? Critics highlight their subjective nature. That is, they are perceived by the experiencer but may not be observable by others. While many of these experiences are subjective, we can observe many of them, such as micro-PK, bio-PK, telepathy, remote viewing, and precognition, with objective methods and in laboratory settings. An important paper demonstrating the validity of these phenomena was published in the *American Psychologist,* the official journal of the American Psychological Association (Cardeña 2018).

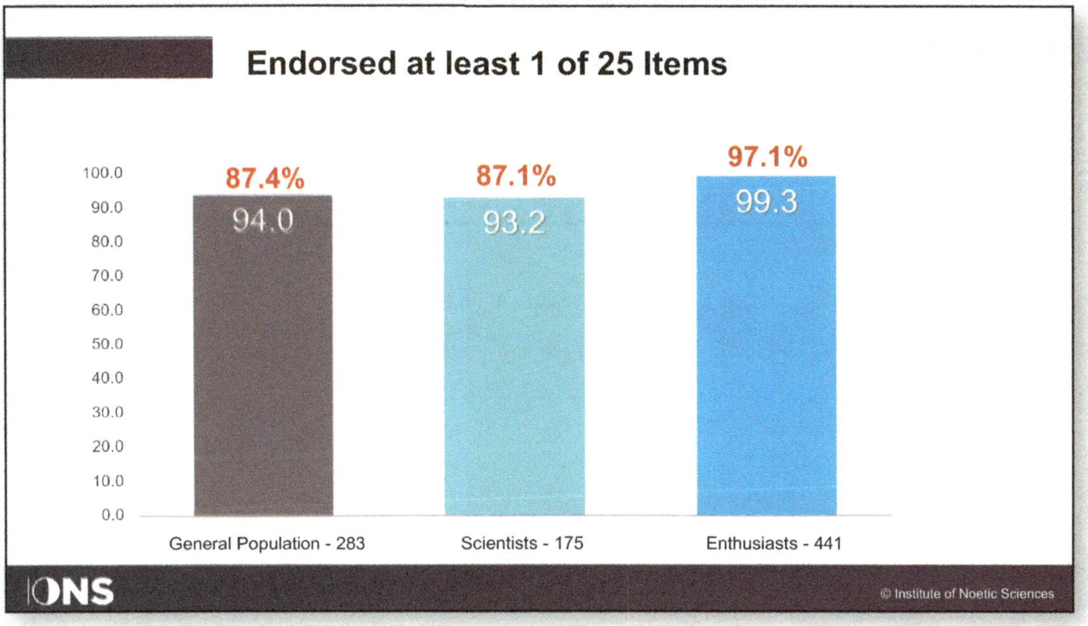

Figure 4

The paper summarizes meta-analyses on various channeling experiences.

This and other papers present a picture of growing evidence that the anomalous phenomenon of psychokinesis is valid. For example, early micro-PK evidence from studies evaluating if our intention could influence what side a rolled die would land on found small but statistically significant effects (ES = 0.0072; Radin and Ferrari 1991). Other evidence for micro-PK using random number generators as the probabilistic target with multiple laboratory studies demonstrated that participants can influence random processes (ES = 0.50003; Bosch et al. 2006). Random number generator studies outside of the laboratory, where the effect of "collective consciousness" is being observed, have also produced significant effect sizes (ES = 0.33; Nelson 2015).

Bio-PK studies show a similar pattern where significant non-zero effects have been demonstrated in various settings. The most applicable to human health are the results we see when positive healing intention is

Remote Influence

Database	*k* (trials)	*Z*	*p*	*ES*	95% CI
Remote influence (adapted from Schmidt, 2015)					
EDA-DMILS (hom)	36		.001	.106	[.04, .17]
Remote staring	15		.013	.128	[.03, .23]
Attention facilitation	11		.029	.114	[.01, .22]

IONS

Figure 5

directed to another person's body (ES values of 0.12–0.26; Roe et al. 2015). We also observe positive results when the intentions are sent to nonhuman animals, cells, or plants (Roe et al. 2015). A meta-analysis of fifty-seven studies found that the effect size for noncontact healing studies was 0.203 for all human studies and 0.224 for the twenty-seven studies of the fifty-seven that were high quality, 0.25 for in-vitro studies, 0.25 for nonhuman animal studies, and 0.2 for plant studies, all of which are statistically significant. Modalities such as Therapeutic Touch, Reiki, and spiritual healing for conditions including pain, cancer, mental health symptoms, and hypertension have had positive research results (Jain et al. 2015; Rao et al. 2016; Yount et al. 2021). In our studies, we have seen hand and wrist pain improvement from a 20-minute energy medicine session in almost 200 participants (Yount et al. 2021) and significant well-being improvements in participants with acute injuries, memory impairment, depression, and anxiety (Wahbeh, Speirn, and Yount 2022). Significant effects have also been observed for remote staring (ES = 0.13) and remote facilitation (ES = 0.11; Schmidt 2015; Sheldrake 2015). In these studies, "receivers" can tell that they are being stared at or helped somehow, like test-taking.

One rigorously replicated bio-PK paradigm with significant findings is the distant mental intention on living systems (DMILS) (ES = 0.11; Schmidt 2015), where a "sender" directs their positive intention towards a "receiver" and researchers simultaneously observe the physiology of both parties. Different physiological measures have changed, such as electrodermal activity, blood volume, fMRI (Achterberg et al. 2005), EEG (T.L. Richards et al. 2005), and heart rate. To date, three meta-analyses examined the results for the DMILS studies and found significant collated results (Schmidt 2012, 2015; Schmidt et al. 2004) (**Figure 5**). Most importantly to the discussion of shifting scientific paradigms, DMILS protocols results demonstrate that the target changes from intention are observed instantaneously and across far distances.

Receptive anomalous phenomena also have evidence demonstrating their validity. For example, knowing mental impressions that one would not

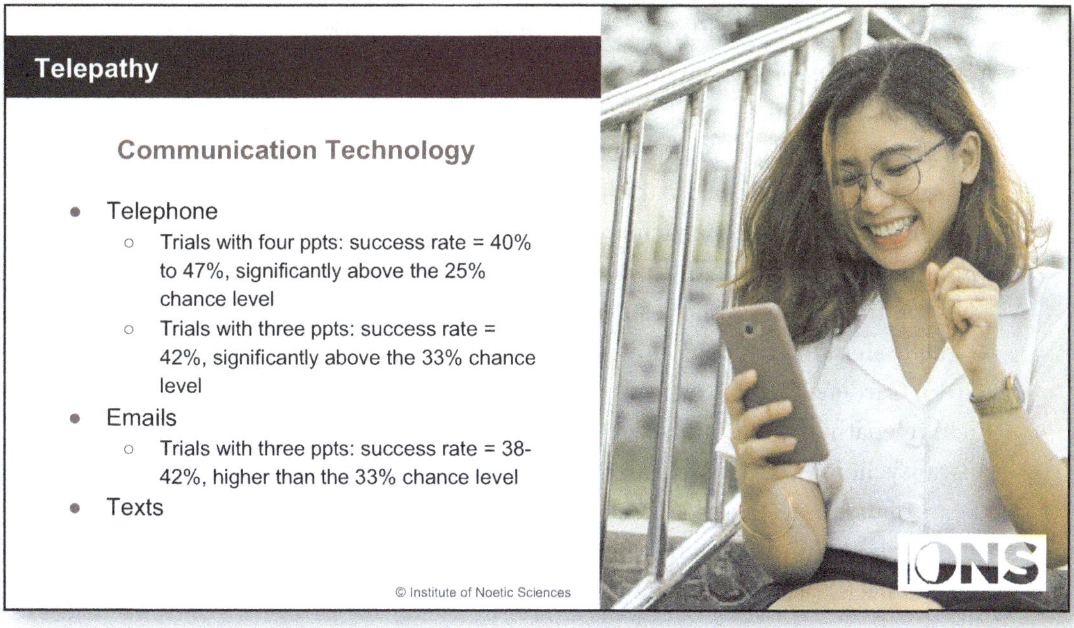

Figure 6

Telepathy - Ganzfeld

Database	*k* (trials)	*Z*	*p*	*ES*	95% CI
	Ganzfeld (adapted from Storm et al., 2010b)				
Combined (all)	108	8.31	$<.10^{-16}$	.142	—
Combined (hom)	102	8.13	$<.10^{-16}$	.135	[.10, .17]
New (all)	30	6.34	1.15×10^{-10}	.152	—
New (hom)	29	5.48	2.13×10^{-8}	.142	[.07, .22]

IONS

Figure 7

usually know, or telepathy, has been studied using various paradigms. One paradigm, telephone telepathy, emulates the real-world experience of knowing who is calling you. People all over the globe have participated in telephone telepathy experiments with above-chance results (Sheldrake 2015) (**Figure 6**). The ganzfeld paradigm proposes that telepathy is easier when sensory input for the participant is moderated by placing translucent ping-pong balls over the eyes, having red light in the space, and white noise playing through headphones. Ganzfeld telepathy experiments have demonstrated significant positive effects (Baptista, Derakhshani, and Tressoldi 2015; Cardeña 2018; Storm, Tressoldi, and Di Risio 2010) (**Figure** 7). Ganzfeld telepathy studies are some of the most reliable and robust findings in parapsychology. Telepathy can also occur during sleep states. More than 25 years of dream telepathy studies and follow-up research show that dreamers dreamt about and correctly identified the targets more often than expected by chance (Storm and Rock 2015; Storm et al. 2017).

Remote Viewing

Database	*k* (trials)	*Z*	*p*	*ES*	95% CI
Remote viewing (adapted from Baptista, Derakhshani, & Tressoldi, 2015; Dunne & Jahn, 2003; Milton, 1997)					
SRI	(770)			.20	[.17, .23]
SAIC	(445)			.23	[.19, .27]
Milton (hom)	75 (2,682)	5.85	2.46×10^{-9}	.17	[.10, .22]
Dunne & Jahn	88 (653)	5.42	3×10^{-8}	.21	[.18, .24]
Bierman & Rabeyron	(550)			.27	[.23, .31]
1994–2014	(314)			.39	[.14, .64]

IONS

Figure 8

Remote viewing is another phenomenon with robust evidence for its validity. Remote viewing meta-analyses report some of the most substantial effects in parapsychology with effect sizes ranging from 0.17 to 0.39 (Baptista et al. 2015; Cardeña 2018; Dunne and Jahn 2003; Milton 1997) (**Figure 8**). Remote viewing studies are especially interesting because we can observe its practical applications in military campaigns (May and Marwaha 2018), stock market, futures or other financial market predictions, sports event outcomes, locations of missing persons or criminal cases, and finding unknown archaeological sites (Harary and Targ 1985; Kolodziejzyk 2013; Schwartz 2019; Schwartz et al. 2019; Smith et al. 2014). These results demonstrate that people can access information from distant locations they have never been to and even distant times in the case of archaeological digs.

Precognition, or knowing the future, has been tested with a variety of laboratory protocols. For example, forced-choice tests, in which the participant chooses one of five cards that they believe will be the target card in the

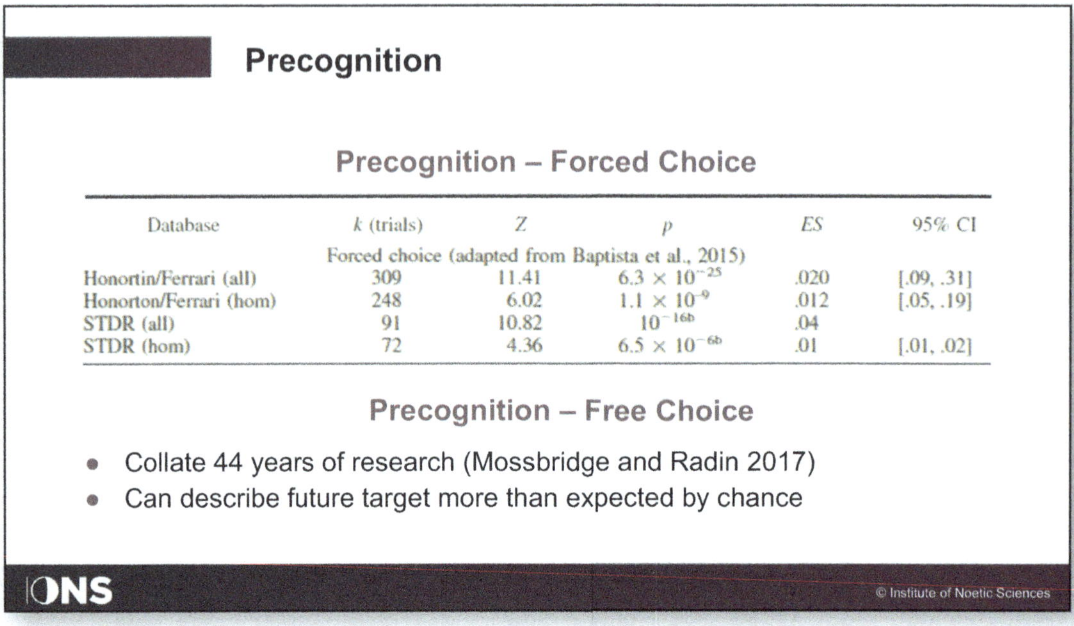

Figure 9

future, have shown significant effects. Free-choice tests, where the person can describe or write out their perceptions, have slightly better effects than the forced-choice protocol. When we combine all of these results, we observe that people can guess what will show up in the future more than they would by chance (Honorton et al. 2018; Mossbridge and Radin 2017; Storm et al. 2012) (**Figure 9**). Using the Bem reverse priming protocol for precognition, the effect sizes range from 0.09–0.11 (Bem et al. 2015).

Interestingly, when we look at unconscious changes in the body about the future (presentiment) rather than conscious perception of the future, we see effects around 0.17–0.24 (**Figure 10**). Presentiment laboratory studies demonstrate that the body reacts to randomly delivered information approximately one to ten seconds in the future (Mossbridge, Tressoldi, and Utts 2012; Mossbridge et al. 2014). These results demonstrate that people can consciously and unconsciously know or feel the future.

This review of studies on psychokinesis, telepathy, remote viewing, and

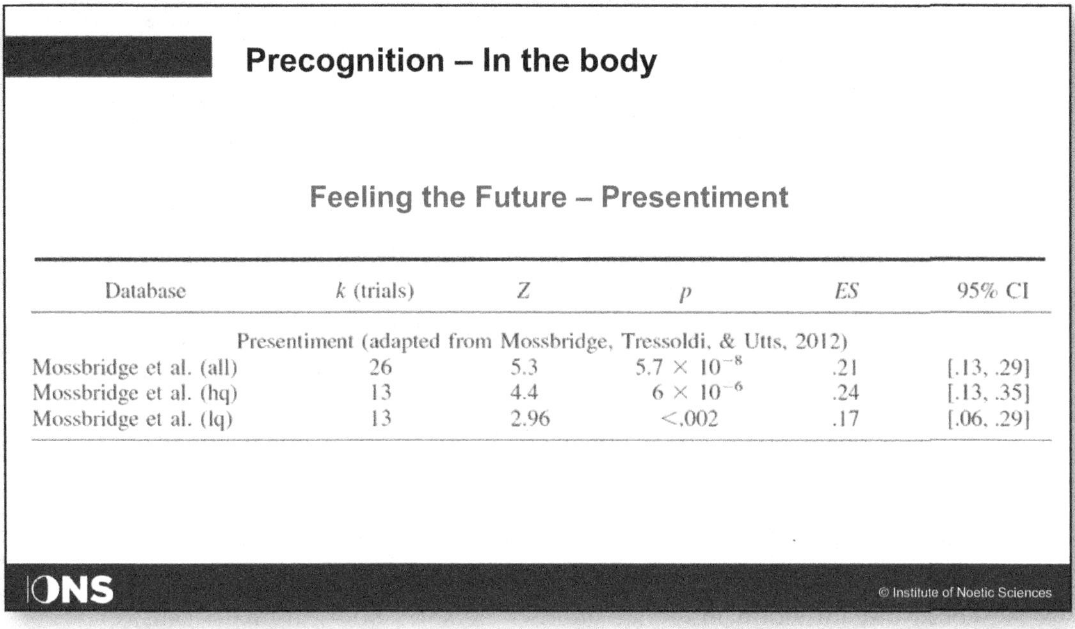

Database	k (trials)	Z	p	ES	95% CI
Presentiment (adapted from Mossbridge, Tressoldi, & Utts, 2012)					
Mossbridge et al. (all)	26	5.3	5.7×10^{-8}	.21	[.13, .29]
Mossbridge et al. (hq)	13	4.4	6×10^{-6}	.24	[.13, .35]
Mossbridge et al. (lq)	13	2.96	<.002	.17	[.06, .29]

Figure 10

precognition is just a brief overview of the voluminous information available about the validity of these phenomena. We also saw that these experiences are pervasive. Still, some might argue that the results we observe in daily life and the laboratory are impossible because they transcend time and space, which contradicts our dominant scientific paradigm. Some deniers would say that the overwhelming evidence I just showed you, that these are natural phenomena, is impossible. Instead, they argue that there is some other—yet to be revealed—explanation that logically describes why they appear to transcend time and space. For example, we recently received this manuscript rejection for a protocol studying these phenomena from *PLOS One* editors.

> After careful consideration and consultation with other staff editors, we have decided that your manuscript does not meet our publication criteria and therefore must be rejected. This study investigates the topic of psi phenomena; however this is based on the premise that psi ability is

> a valid and reproducible phenomenon. As the experimental and scientific validity of such phenomena have not been demonstrated nor initially established, we feel that this study lacks scientific relevance and would not contribute towards academic knowledge (PLOS, personal communication, March 8, 2022).

This reply shows an apparent dissonance between our dominant paradigm understanding of the world and our experimental evidence. Our dominant current scientific paradigm is materialism or physicalism, and our current model for how the world works is based on a materialistic paradigm. Materialism promotes the view that nothing exists except matter and its movements and changes. In materialism, affecting something at a distance would depend on physical laws that the material world is constrained by: time flows in a linear, forward direction, and traveling distances in space requires time. We do not see that in distant intention experiments. How could our intention influence another person's body, animals, cells, or machines at a distance and instantaneously?

Similarly, only the sensory input from our traditional five senses or information we sense from tools we use would allow us to gain information. How can we perceive information from somewhere far distant, the future, or another's mind? According to materialism, these effects we observe in the lab are impossible. While materialism explains much about how our physical world works, it does not explain everything, including these anomalous phenomena. Thus, we need to shift the scientific paradigm and embrace a new model of the nature of reality that allows for these phenomena.

We need to shift beyond materialism and into postmaterialism. In postmaterialism, we acknowledge that we are all interconnected. From cosmology, we know that there was a point in time when everything in the Universe was connected (Buniy and Hsu 2012) (**Figure 11**). In theory, this means that the physical properties of a photon at one end of the Universe are entirely linked to those of another photon at the other end of the Universe. The physical properties of the photons, like position, speed, and spin, are connected. Scientists have repeatedly

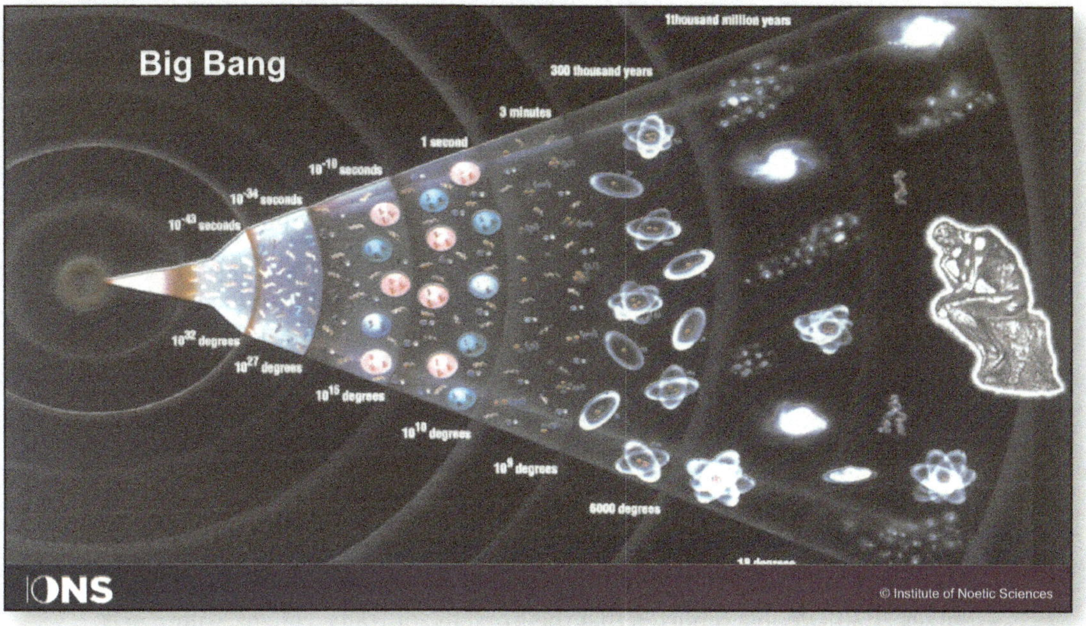

Figure 11

demonstrated this quantum entanglement with photons, small molecules, and even macroscopic diamonds (Bub 2019). Laboratories have confirmed quantum entanglement with up to eight photons (Yao et al. 2012). Researchers also observe quantum entanglement in photon pairs from two distant, ancient quasars emitted 7.8 billion years ago and 12.2 billion years ago (Rauch et al. 2018).

If everything is interconnected, then the idea that we might know information from a distant time or space does not seem so far-fetched. In fact, many academics from multiple fields are proposing alternate models: ones in which this interconnectedness is consciousness, meaning that consciousness is fundamental; or other alternative or complementary models of reality that allow for these phenomena (Chalmers 1996; Cook 2020; Currivan 2017; Faggin 2021; Fields et al. 2018; Goff 2019; Kastrup 2017; Prakash et al. 2020). If we assume that consciousness is fundamental, then nonlocal consciousness phenomena, where our consciousness transcends our conventional notions of time and space, make sense.

A paradigm shift is required to encompass these anomalous phenomena, considering their commonality and the evidence demonstrating their validity in the laboratory and real life. That shift is happening. Perhaps holding the hypothetical assumption that consciousness is fundamental and focusing on what we can learn about the mechanism, mediators, moderators, and practical applications of nonlocal consciousness will help us jump across the paradigm chasm we are straddling.

Regardless of prevalence, validity, and models of reality, one often asks why it is important to study such phenomena. Why does it matter? Well, most people say that their various channeling experiences are beneficial, inspirational, and positively impact their lives (Griffiths et al. 2008; Kennedy and Kanthamani 1995a; Ellison and Fan 2008; Wahbeh et al. 2018; Richards 1991). Trance channeling and mediumship positively impact the channeler's life (Negro, Palladino-Negro, and Louzã 2002; Moreira-Almeida and Cardeña 2011; Wahbeh, Carpenter, and Radin 2018; Wahbeh et al. 2019; Wahbeh and Butzer 2020). People reported that their psychic or transcendent experiences were valuable in one study, and very few participants rated the psychic experiences as harmful (Kennedy and Kanthamani 1995b). If humans could access information not previously known to them that could support our world, then why not? If we could use our positive intentions to heal without side effects, why not?

In conclusion, the anomalous phenomena described in the title of this talk are, in actuality, not anomalous but typical and expected. The phenomena are valid in that we can observe them with objective measures in the laboratory with non-zero effects. The fact that these experiences transcend time and space and the notion that our consciousness is housed in our physical brain is contradictory to a materialist scientific paradigm. While materialism explains much in our world, it does not explain everything, including these phenomena. Physicists, cosmologists, philosophers, and other academics are working hard to develop new theories to help describe a postmaterialist world. ◊

References

Achterberg, J., K. Cooke, T. Richards, L.J. Standish, and L. Kozak. 2005. "Evidence for Correlations Between Distant Intentionality and Brain Function in Recipients: A Functional Magnetic Resonance Imaging Analysis." *Journal of Alternative and Complementary Medicine* 11 (6): 965–971.

Baptista, J., M. Derakhshani, and P.E. Tressoldi. 2015. "Explicit anomalous cognition: A review of the best evidence in ganzfeld, forced choice, remote viewing and dream studies." In *Parapsychology: A Handbook for the 21st Century,* edited by E. Cardeña, J. Palmer, and D. Marcusson-Clavertz, 192–214. Jefferson, NC: McFarland & Company.

Bem, D.J., P. Tressoldi, T. Rabeyron, and M. Duggan. 2015. "Feeling the future: A meta-analysis of 90 experiments on the anomalous anticipation of random future events [version 2; peer review: 2 approved]" *F1000Research* 4: 1188. https://f1000research.com/articles/4-1188/v2

Bosch, H., F. Steinkamp, and E. Boller. 2006. "Examining psychokinesis: The interaction of human intention with random number generators—A meta-analysis." *Psychological Bulletin* 132 (4): 497–523. https://psycnet.apa.org/doi/10.1037/0033-2909.132.4.497

Bourguignon, E. 1976. *Possession.* San Francisco, CA: Chandler & Sharp Publishers.

Bub, J. 2019. "Quantum Entanglement and Information." In *The Stanford Encyclopedia of Philosophy* (Spring 2019 Edition), edited by E.N. Zalta. Stanford, CA: The Metaphysics Research Lab, Stanford University. Accessed on July 7, 2022. https://plato.stanford.edu/archives/spr2019/entries/qt-entangle/

Buniy, R.V., and S.D.H. Hsu. 2012. "Everything is entangled." *Physics Letters B* 718 (2): 233–236. https://doi.org/10.1016/j.physletb.2012.09.047

Cardeña, E. 2018. "The experimental evidence for parapsychological phenomena: A review." *American Psychologist* 73 (5): 663–677. https://doi.org/10.1037/amp0000236

Castro, M., R. Burrows, and R. Wooffitt. 2014. "The Paranormal is (Still) Normal: The Sociological Implications of a Survey of Paranormal Experiences in Great Britain." *Sociological Research Online* 19 (3): 16. https://doi.org/10.5153/sro.3355

Chalmers, D.J. 1996. *The Conscious Mind: In Search of a Fundamental Theory.* New York, NY: Oxford University Press.

Cohn, S.A. 1994. "A Survey on Scottish Second Sight." *Journal of the Society for Psychical Research* 59 (835): 385–400. Accessed on July 7, 2022. https://era.ed.ac.uk/handle/1842/9674?show=full

Cook, G. 2020. "Does Consciousness Pervade the Universe?" *Scientific American.* Accessed on July 7, 2022. https://www.scientificamerican.com/article/does-consciousness-pervade-the-universe/

Currivan, J. 2017. *The Cosmic Hologram: In-formation at the Center of Creation.* Rochester, VT: Inner Traditions.

Dunne, B.J., and R.G. Jahn. 2003. "Information and Uncertainty in Remote Perception Research." *EXPLORE* 3 (3): 254–269. https://doi.org/10.1016/j.explore.2007.03.010

Ellison, C.G., and D. Fan. 2008. "Daily Spiritual Experiences and Psychological Well-being Among US Adults." *Social Indicators Research* 88 (2): 247–271. https://doi.org/10.1007/s11205-007-9187-2

Faggin, F. 2021. "Consciousness Comes First." In *Consciousness Unbound: Liberating Mind from the Tyranny of Materialism,* edited by E.F. Kelley and P. Marshall, 283–322. Lanham, MD: Rowman & Littlefield Publishers.

Fields, C., D.D. Hoffman, C. Prakash, and M. Singh. 2018. "Conscious agent networks: Formal analysis and application to cognition." *Cognitive Systems Research* 47: 186–213. https://doi.org/10.1016/j.cogsys.2017.10.003

Goff, P. 2019. *Galileo's Error: Foundations for a New Science of Consciousness.* New York City, NY: Pantheon Books.

Greeley, A. 1987. "Mysticism Goes Mainstream." *American Health* 7: 47–49.

Griffiths, R.R., W.A. Richards, M.W. Johnson, U.D. McCann, and R. Jesse. 2008. "Mystical-type experiences occasioned by psilocybin mediate the attribution of personal meaning and spiritual significance 14 months later." *Journal of Psychopharmacology* 22 (6): 621–632. https://doi.org/10.1177/0269881108094300

Haraldsson, E. 1985. "Representative national surveys of psychic phenomena: Iceland, Great Britain, Sweden, USA and Gallup's multinational survey." *Journal of the Society for Psychical Research* 53 (801): 145–158.

———. 2011. "Psychic experiences a third of a century apart: Two representative surveys in Iceland with an international comparison." *Journal of the Society for Psychical Research* 75 (903): 76.

———, and J.M. Houtkooper. 1991. "Psychic experiences in the multinational human values study: Who reports them?" *Journal of the American Society for Psychical Research* 85 (2): 145–165.

Harary, K., and R. Targ. 1985. "A new approach to forecasting commodity futures." *PSI Research* 4 (3-4): 79–88.

Honorton, C., D.B. Ferrari, and G. Hansen. 2018. "'Future telling': Meta-analysis of Forced-Choice Precognition Experiments, 1935–1987." *Journal of Parapsychology* 53 (4): 281–308.

Jain, S., R. Hammerschlag, P. Mills, L. Cohen, R. Krieger, C. Vieten, and S. Lutgendorf. 2015. "Clinical Studies of Biofield Therapies: Summary, Methodological Challenges, and Recommendations." *Global Advances in Health and Medicine* 4 (1_Suppl): 58–66. https://doi.org/10.7453/gahmj.2015.034.suppl

Kastrup, B. 2017. "Self-transcendence Correlates with Brain Function Impairment." *Journal of Cognition and Neuroethics* 4: 33–42.

Kennedy, J.E., and H. Kanthamani, H. 1995a. "An exploratory study of the effects of paranormal and spiritual experiences on peoples' lives and well-being." *Journal of the American Society for Psychical Research* 89 (3): 249–264.

———. 1995b. "Association between anomalous experiences and artistic creativity and spirituality." *Journal of the American Society for Psychical Research* 89: 333–343.

Kolodziejzyk, G. 2013. "Greg Kolodziejzyk's 13-year associative remote viewing experiment results." *Journal of Parapsychology* 76: 349–368.

May, E.C., and S.B. Marwaha, eds. 2018. *The Star Gate Archives: Reports of the United States Government Sponsored Psi Program, 1972-1995 Volume 1: Remote Viewing, 1972-1984,* Illustrated Edition. Jefferson, NC: McFarland & Company.

McClenon, J. 1993. "Surveys of Anomalous Experience in Chinese, Japanese, and American Samples." *Sociology of Religion* 54 (3): 295–302. https://doi.org/10.2307/3711723

Milton, J. 1997. "Meta-Analysis of Free-Response ESP Studies Without Altered States of Consciousness." *Journal of Parapsychology* 61: 279–319.

Moreira-Almeida, A., and E. Cardeña. 2011. "Differential diagnosis between non-pathological psychotic and spiritual experiences and mental disorders: A contribution from Latin American studies to the ICD-11." *Brazilian Journal of Psychiatry* 33 (Suppl 1). https://doi.org/10.1590/S1516-44462011000500004

Mossbridge, J.A., and D. Radin. 2018. "Precognition as a form of prospection: A review of the evidence." *Psychology of Consciousness: Theory, Research, and Practice* 5 (1): 78–93. https://psycnet.apa.org/doi/10.1037/cns0000121

———, P. Tressoldi, and J. Utts. 2012. "Predictive physiological anticipation preceding seemingly unpredictable stimuli: A meta-analysis." *Frontiers in Psychology* 3: 390. https://doi.org/10.3389/fpsyg.2012.00390

———, P. Tressoldi, J. Utts, J.A. Ives, D. Radin, and W.B. Jonas. 2014. "Predicting the unpredictable: Critical analysis and practical implications of predictive anticipatory activity." *Frontiers in Human Neuroscience* 8: 146. https://doi.org/10.3389/fnhum.2014.00146

Negro Jr, P.J., P. Palladino-Negro, and M. R. Louzã. 2002. "Do Religious Mediumship Dissociative Experiences Conform to the Sociocognitive Theory of Dissociation?" *Journal of Trauma & Dissociation* 3 (1): 51–73. https://www.tandfonline.com/doi/abs/10.1300/J229v03n01_05

Nelson, R. D. 2015. "Implicit Physical Psi: The Global Consciousness Project." In *Parapsychology: A Handbook for the 21st Century,* edited by E. Cardena, J. Palmer, and D. Marcusson-Clavertz, 282–292. Jefferson, NC: McFarland & Company.

Palmer, J. 1979. "A community mail survey of psychic experiences." *Journal of the American Society for Psychical Research* 73 (3): 221–251.

Prakash, C., C. Fields, D.D. Hoffman, R. Prentner, and M. Singh. 2020. "Fact, Fiction, and Fitness." *Entropy* 22 (5): 514. https://doi.org/10.3390/e22050514

Radin, D., and D. Ferrari. 1991. "Effects of Consciousness on the Fall of Dice: A Meta-Analysis." *Journal of Scientific Exploration* 5 (1): 61–83.

Rao, A., L.D. Hickman, D. Sibbritt, P.J. Newton, and J.L. Phillips. 2016. "Is energy healing an effective non-pharmacological therapy for improving symptom management of chronic illnesses? A systematic review." *Complementary Therapies in Clinical Practice* 25: 26–41. https://doi.org/10.1016/j.ctcp.2016.07.003

Rauch, D., J. Handsteiner, A. Hochrainer, J. Gallicchio, A.S. Friedman, C. Leung, B. Liu, L. Bulla, S. Ecker, F. Steinlechner, R. Ursin, B. Hu, D. Leon, C. Benn, A. Ghedina, M. Cecconi, A.H. Guth, D.I. Kaiser, T. Scheidl, and A. Zeilinger. 2018. "Cosmic Bell Test Using Random Measurement Settings from High-Redshift Quasars." *Physical Review Letters* 121 (8): 080403. https://doi.org/10.1103/PhysRevLett.121.080403

Richards, D.G. 1991. "A study of the correlations between subjective psychic experiences and dissociative experiences." *Dissociation: Progress in the Dissociative Disorders* 4 (2): 83–91.

Richards, T.L., L. Kozak, L.C. Johnson, and L.J. Standish. 2005. "Replicable Functional Magnetic Resonance Imaging. Evidence of Correlated Brain Signals Between Physically and Sensory Isolated Subjects." *Journal of Alternative and Complementary Medicine* 11 (6): 955–963. https://doi.org/10.1089/acm.2005.11.955

Roe, C.A., C. Sonnex, and E.C. Roxburgh. 2015. "Two Meta-Analyses of Noncontact Healing Studies." *EXPLORE: The Journal of Science and Healing* 11 (1): 11–23. https://doi.org/10.1016/j.explore.2014.10.001

Ross, C.A., and S. Joshi. 1992. "Paranormal experiences in the general population." *The Journal of Nervous and Mental Disease* 180 (6): 357–361. https://doi.org/10.1097/00005053-199206000-00004

Schmidt, S. 2012. "Can We Help Just by Good Intentions? A Meta-Analysis of Experiments on Distant Intention Effects." *The Journal of Alternative and Complementary Medicine* 18 (6): 529–533. https://doi.org/10.1089/acm.2011.0321

———. 2015. "Experimental research on distant intention phenomena." In *Parapsychology: A Handbook for the 21st Century,* edited by E. Cardeña, J. Palmer, and D. Marcusson-Clavertz, 244–257. Jefferson, NC: McFarland & Company.

———, R. Schneider, J. Utts, and H. Walach. 2004. "Distant intentionality and the feeling of being stared at: Two meta-analyses." *British Journal of Psychology* 95 (2): 235–247. https://doi.org/10.1348/000712604773952449

Schwartz, S.A. 2019. "The Location and Reconstruction of a Byzantine Structure in Marea, Egypt, Including a Comparison of Electronic Remote Sensing and Remote Viewing." *Journal of Scientific Exploration* 33 (3): 451–480. Retrieved from https://journalofscientificexploration.org/index.php/jse/article/view/1479

———, R.J. De Mattei, and R.C. Smith. 2019. "The Caravel Project: The Location, Description, and Reconstruction of Marine Sites Through Remote Viewing, Including Comparison With Aerial Photography, Geological Coring, and Electronic Remote Sensing." *Zeitschrift Für Anomalistik* 19: 113–139. Retrieved from: https://stephanaschwartz.com/wp-content/uploads/2010/03/caravel.pdf

Sheldrake, R. 2015. "Psi in Everyday Life." In *Parapsychology: A Handbook for the 21st Century,* edited by E. Cardeña, J. Palmer, and D. Marcusson-Clavertz, 350. Jefferson, NC: McFarland & Company.

Smith, C.C., D. Laham, and J. Moddel. 2014. "Stock Market Prediction Using Associative Remote Viewing by Inexperienced Remote Viewers." *Journal of Scientific Exploration* 28 (1): 7–16.

Storm, L., and A.J. Rock. 2015. "Dreaming of Psi: A Narrative Review and Meta-Analysis of Dream-ESP Studies at the Maimonides Dream Laboratory and Beyond." In *Stanley Krippner: A Life of Dreams, Myths, and Visions. Essays on His Contributions and Influence,* edited by J.A. Davies and D.B. Pitchford, 117–138. Colorado Springs, CO: University Professors Press.

———, P.E. Tressoldi, and L. Di Risio. 2010. "Meta-analysis of free-response studies, 1992-2008: Assessing the noise reduction model in parapsychology." *Psychological Bulletin* 136 (4): 471–485. https://doi.org/10.1037/a0019457

———, P.E. Tressoldi, and L. Di Risio. 2012. "Meta-analysis of ESP studies, 1987-2010: Assessing the success of the forced-choice design in parapsychology." *Journal of Parapsychology* 76 (2): 243–273.

———, S.J. Sherwood, C.A. Roe, P.E. Tressoldi, A.J. Rock, and L. Di Risio. 2017. "On the correspondence between dream content and target material under laboratory conditions: A meta-analysis of dream-ESP studies, 1966-2016." *International Journal of Dream Research* 10 (2): 120–140. https://psycnet.apa.org/record/2018-14483-003

Thalbourne, M.A. 2003. *A Glossary of Terms Used in Parapsychology.* Charlottesville, VA: Puente Publications.

Vieten, C., H. Wahbeh, B.R. Cahn, K. MacLean, M. Estrada, P. Mills, M. Murphy, S. Shapiro, D. Radin, Z. Josipovic, D.E. Presti, M. Sapiro, J.C. Bays, P. Russell, D. Vago, F. Travis, R. Walsh, and A. Delorme. 2018. "Future directions in meditation research: Recommendations for expanding the field of contemplative science." *PLOS ONE* 13 (11): e0205740. https://doi.org/10.1371/journal.pone.0205740

Wahbeh, H., and B. Butzer. 2020. "Characteristics of English-speaking trance channelers." *EXPLORE* 16 (5): 304–309. https://doi.org/10.1016/j.explore.2020.02.002

———, L. Carpenter, and D. Radin. 2018. "A mixed methods phenomenological and exploratory study of channeling." *Journal of the Society for Psychical Research* 82 (3): 129–148.

———, N. Fry, and P. Speirn. 2022. "The Noetic Signature Inventory: Development and Validation." *PsyArXiv.* https://doi.org/10.3389/fpsyg.2022.838582

———, P. Speirn, and G. Yount. 2022. "Extended Perception Corroboration: A Pilot Study with Energy Medicine Practitioners." *PsyArXiv.* https://doi.org/10.31234/osf.io/ukygq

———, C. Cannard, J. Okonsky, and A. Delorme. 2019. "A physiological examination of perceived incorporation during trance [version 2; peer review: 2 approved]." *F1000Research* 8 (67). https://doi.org/10.12688/f1000research.17157.2

———, D. Radin, J. Mossbridge, C. Vieten, and A. Delorme. 2018. "Exceptional experiences reported by scientists and engineers." *Explore* 14 (5): 329–341. https://doi.org/10.1016/j.explore.2018.05.002

———, N. Fry, P. Speirn, L. Hrnjic, E. Ancel, and E. Niebauer. 2022. "Qualitative analysis of first-person accounts of noetic experiences [version 3; peer review: 2 approved]." *F1000Research* 10: 497. https://doi.org/10.12688/f1000research.52957.3

Yao, X.-C., T.-X. Wang, P. Xu, H. Lu, G.-S. Pan, X.-H. Bao, C.-Z. Peng, C.-Y. Lu, Y.-A. Chen, and J.-W. Pan. 2012. "Observation of eight-photon entanglement." *Nature Photonics* 6 (4): 225–228. https://doi.org/10.1038/nphoton.2011.354

Yount, G., A. Delorme, D. Radin, L. Carpenter, K. Rachlin, J. Anastasia, M. Pierson, S. Steele, H. Mandell, A. Chagnon, and H. Wahbeh. 2021. "Energy Medicine treatments for hand and wrist pain: A pilot study." *EXPLORE* 17 (1): 11–21. https://doi.org/10.1016/j.explore.2020.10.015

Session 1 | First Commentary
on the Presentation by Dr. Helané Wahbeh

The Source of Channeled Information and the Survival of Consciousness

Professor Masayuki Ohkado
Graduate School of Global Humanics,
Chubu University, Japan

I have been working on what are called "anomalous phenomena" for over a decade, especially on cases such as children claiming to have past-life memories and near-death experiences.

Dr. Wahbeh's presentation has elegantly summarized the bulk of the literature on "anomalous phenomena" that have been conducted over the past 150 years. I totally agree with her conclusion that academics should work hard to "develop new theories to help describe a postmaterialist world."

Actually, this conclusion is explicitly stated in "The Manifesto for a Post-Materialist Science," published by Mario Beauregard and others in 2014, and it was immediately signed by many scientists, professors, MDs, and thought

About the Speaker

- Professor, Department of Language and Cultures, Graduate School of Global Humanics, Chubu University, Japan
- Visiting Professor, Division of Perceptual Studies, University of Virginia, USA
- Director, Society for Mind-Body Science (Since 2012)
- Research in near-death experiences, past-life memories, prenatal memories, postnatal memories, and reincarnation
- PhD, Liberal Arts, University of Amsterdam, Netherlands

leaders. As of March 2022, the number of signatories exceeds 400.

In this commentary, I would like to supplement Dr. Wahbeh's presentation by talking about "anomalous phenomena" not discussed or focused on in her presentation.

I have read Dr. Wahbeh's great book, *The Science of Channeling,* and I know that she is well aware of the issue I would like to comment on. But I think it is worth mentioning here, for two reasons: 1) not everyone may be familiar with her book, and 2) it is one of the central concerns for human beings, especially those who do not have personal channeling experiences like Dr. Wahbeh.

The issue is about the source of information you channel with, especially when you obtain information about a deceased person. There are at least three possibilities. (Dr. Wahbeh gives four in her book.) The first is people who have known the deceased, or records, such as documents, about the deceased. You may use your channeling ability and tap into the minds of those who have known the deceased person, or consult records about the person, and receive the relevant information. The second is the so-called "psychic reservoir" or "akashic records": a repository of every thought, word, and deed of every living being, in all of time—past, present, and future. The third is the spirit of the deceased person.

The second and third possibilities guarantee that death is not an annihilation. For those who are afraid of death and those who are suffering from the death of a loved one, the third possibility would be the most welcome one.

I do not have time to deal with the detailed arguments about which of the three is the source of information in a given example, and I will just say that there is ample evidence suggesting that the third possibility is a likely option. In other words, we have some evidence that human consciousness can survive bodily death.

The "anomalous phenomena" pertinent to the discussion of the survival of consciousness but not focused on in Dr. Wahbeh's presentation are near-death experiences and past-life experiences, briefly explained below.

Near-death experiences have been intensively studied for more than 40 years and have been shown to have the following basic pattern:

- An out-of-body experience
- A tunnel experience
- Encountering "Beings of Light"
- Encountering deceased people
- Having a life review
- Approaching a border
- Coming back to one's body

The person in question has these as "personal experiences." It would thus be misleading to label them as "getting information" or "channeling." Yet, they are very important "anomalous phenomena."

Past-life experiences are also similar in character to that of near-death experiences in that the experiencer has them as personal experiences.

Largely, past-life experiences can be divided into two types: those experienced by adults, mostly under hypnotic past-life regression, and those experienced by children. In this commentary, I will just focus on children claiming to have past-life experiences.

The Division of Perceptual Studies at the University of Virginia is the center of the study of such children. Over 50 years, researchers affiliated with the division have collected more than 2,600 cases from all over the world and have shown that the phenomena have an interesting pattern:

- At the young age of 2 to 4, children start talking about their past-life memories.
- They show emotions and behaviors appropriate to their memories, but not to their present situations.
- They might have birthmarks or birth defects apparently due to their past-life experiences.
- The people the children talk about are often found.
- As they grow, they talk about their memories less and less, and usually stop talking by the time they become adults.

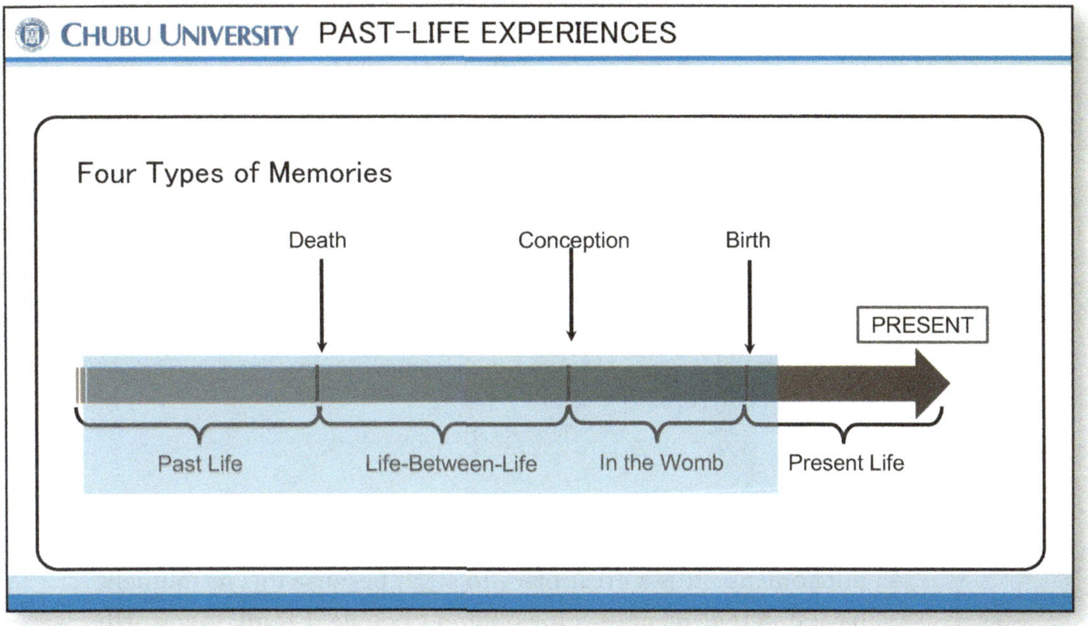

Figure 1

An important fact about these children is that some of them have memories after they died in their past lives: memories between their deaths and conception, memories of being in the womb, and memories of being born. In other words, some of them have past-life memories as those separate from but continuous with their present lives (**Figure 1**). In such cases, it is almost inconceivable to say that they obtained these memories through channeling, just as it is inconceivable to say that all our memories in our present lives are obtained through channeling. Also, experiencers say that what they claim to remember is what they have experienced.

In conclusion, I think Dr. Wahbeh's approach to a group of "anomalous phenomena" by calling them "channeling" is acceptable, but an important set of phenomena related to the argument about the survival of human consciousness should also be dealt with. ◊

References

Beauregard, M., G.E. Schwartz, L. Miller, L. Dossey, A. Moreira-Almeida, M. Schlitz, R. Sheldrake, and C. Tart. 2014. "Manifesto for a Post-Materialist Science." *EXPLORE* 10 (5): 272–274. https://doi.org/10.1016/j.explore.2014.06.008

Session 1 | Second Commentary on the Presentation by Dr. Helané Wahbeh

Anomalous Phenomena that Require Shifts in the Scientific Paradigm

Professor Arnaud Delorme
CerCo Laboratory, CNRS,
Paul Sabatier University, Toulouse, France

Dr. Wahbeh started by describing what she meant by "anomalous phenomena." It is a great place to start, because this term might have different meanings for different people. I think it is worth emphasizing this point, and what "anomalous phenomena" means with respect to current scientific models. Current scientific models have little room for such phenomena. However, a model is based on a set of rules we imagine to describe nature, and these rules may evolve over time. For exam-

About the Speaker

- Professor, CerCo Laboratory, CNRS, Paul Sabatier University, Toulouse, France
- Faculty, Swartz Center for Computational Neuroscience, University of California, San Diego, USA
- Scientist, Institute of Noetic Sciences, Petaluma, California
- Co-Principal Investigator, EEGLAB: Software for Analysis of Human Brain Dynamics
- Primary research interest is in analysis and modeling of human consciousness as captured by high-dimensional electroencephalography (EEG) and other imaging modalities
- PhD in Computational Neuroscience; Investigating how technology could be used to assist meditation practice

ple, the Newtonian paradigm was replaced by Einstein's relativity about 100 years ago. From a Newtonian perspective, some observations returned anomalous results, and Einstein's relativity model was able to explain these results.

I believe the type of anomalous phenomena Dr. Wahbeh describes are of the same nature. Assuming the observations are robust and reproducible, we need to develop new models of nature to explain how they fit with the existing models of physics, in the same way Einstein developed a new model to explain anomalous astronomical measurements.

Dr. Wahbeh then argues that these phenomena, particularly channeling experiences, should not be called "anomalous" because they are well established, with 90 % of people claiming to have experienced them at least once. She also mentions numerous publications in the field, in particular, a recent review by Cardeña (2018). I agree that the term anomalous is misleading. It defines these phenomena with respect to models of nature that we know are incomplete. For example, we know that general relativity and quantum mechanics are not mutually consistent. The term "unexplained" still puts the emphasis on physical models, but "unexplained" is better than "anomalous," as the latter seems to imply that there is something erroneous about the phenomena.

Dr. Wahbeh then argues that although most of the unexplained phenomena reported sprout from subjective experience, objective and quantifiable measurements can often be made. She then proceeds to describe scientific results obtained on this subject. Dr. Wahbeh describes different types of unexplained phenomena, ranging from telepathy, mind-matter interactions, and channeling. She presents statistical analyses of objective measurements in controlled experiments. She also focuses on effect size, an important concept in scientific discovery, as small effect size, even when significant, often points toward artifactual results. She also presents the results of multiple, independently conducted studies for each topic. This approach is commendable. It is difficult to make scientific inferences based on subjective experiences, so Dr. Wahbeh is correct in emphasizing tangible measurements.

Dr. Wahbeh then mentions the stigma and censorship associated with these fringe topics by the scientific community. This censorship seems to sprout from the belief of scientists themselves and from monetary considerations: a scientific journal's reputation is highly correlated with its revenues. Publishers do not want to risk the reputation of their journal by publishing studies on fringe topics, even when they are conducted rigorously.

Dr. Wahbeh finally argues that we need to shift from the materialist paradigm to be able to study these phenomena, and I wholeheartedly agree. The current scientific paradigm is materialist in nature. Materialism is a scientific dead end. It likely cannot account for subjective experience. It is also based on more than a century-old premises that have been proven to be erroneous. For example, according to our best physical models, solid matter does not exist: matter only exists probabilistically, according to quantum mechanics. Does this require a shift in the scientific paradigm? Can we not just continue to use the scientific paradigm to study unexplained phenomena? I think yes and no, and based on her presentation, I would think Dr. Wahbeh agrees.

The scientific paradigm is based on the premise that nature obeys a set of rules. This does not prevent us from studying unexplained phenomena. However, it also assumes that these rules are timeless, deriving from complex symmetries. For example, forces in nature derive from symmetry, as described by Noether's theorem (1918). However, the initial symmetry breaking cannot be explained by science. If there was a rule to explain it, there should be another rule to explain why this initial rule is present in the first place. I think this is where science can meet spirituality. The initial symmetry breaking could very well have happened in consciousness. It links back to Dr. Wahbeh's comment of everything being interconnected, including human consciousness, and points toward a model on integrating consciousness into the scientific paradigm and shifting the scientific paradigm. ◊

References

Cardeña, E. 2018. "The experimental evidence for parapsychological phenomena: A review." *American Psychologist* 73 (5): 663–677. https://psycnet.apa.org/doi/10.1037/amp0000236

Noether, E. 1918. "Invariante Variationsprobleme." *Nachrichten von der Gesellschaft der Wissenschaften zu Göttingen, Mathematisch-Physikalische Klasse 1918:* 235–257. Accessed on July 8, 2022. https://eudml.org/doc/59024

Session 1

Discussion

Menas Kafatos (Session Chair): I would again like to thank the main speakers, presenters, and commentators. I think we are ready to go. There is a question from Didier Guignard. Please, go ahead.

Didier Guignard: My question is for the first speaker, Prof. Marjorie Woollacott. She explained about filtering in the brain, and I see those filters as necessary. My experience in music, such as in playing the piano, is that you need to concentrate on certain aspects. It is thus necessary to have some filters. She also explained that we can remove these filters through meditation. As I see it, meditation also introduces maybe a different filter, because it is a conscious filter.

But it is also very important to do good meditation, in my opinion. I would like to ask how Prof. Woollacott feels about that. Is it necessary to isolate yourself from the environment, from the external noise, so that you can get the internal signals from your inner voice? Dr. Wahbeh spoke about the five external senses, but we also have internal senses. For me, meditation is a way to slightly cut off the external senses and turn more to the five internal senses. In this way, we can go beyond space and time and connect with the whole of the universe.

Marjorie Woollacott: I heard your question, and I thank you very much. I think we are probably in general agreement, because what you are saying is that yes, those filters are very important in our daily life. As I showed with the thalamo-cortical loop, you want to be sure you can turn off, for example, external sounds from your vision when you want to really be hearing someone. And you can also turn down both vision and hearing when you want to meditate. So that is absolutely true. Regarding meditation, I think what we are doing is, we are actually turning down our filters, which are modulating our internal narrative. This allows us to hear what I would call the "internal intuitive sense" and connect with the whole of the universe. Then we can be in a total place of pure inner stillness that you were talking about. My own sense is that, yes, meditation is very important for daily life.

Yet, it is absolutely important for being able to go beyond the distractions of the external world and the mind. It is also important to hear things perhaps from this nonlocal consciousness from other realms and other people within the universe that actually give us information not readily available through the five senses, which I think is what Helané Wahbeh was also talking about. When we lower down those filters, then we have access to these precognitive experiences and other experiences that some people call "anomalous," but they are actually very natural when those filters are turned down.

Kafatos: Very good. Our next question is from Jin Choon Kim.

Jin Choon Kim: My question is for Dr. Woollacott. To understand the postmaterialist era, I think we first need to understand or define the meaning of postmaterialism. In my understanding, there may be three possibilities. The first possibility is non-conventional materialism, not such materials as quarks, leptons, atoms, molecules, etc. of the Standard Model in physics. In non-materialism, you may consider having nonmaterial forms such as information energy fields.

The second possibility is the existence of consciousness or a mind, including mindfulness and psychological topics. The third possibility is the existence of a spiritual or spirit energy, such as in channeling phenomena as explained by Dr. Wahbeh. My question is the following: In your perspective, do you understand postmaterialism as a combination of those three, or one of those three, or two of the three?

Woollacott: Yes, there is one possibility, often referred to as a nondual perspective, which states that consciousness is fundamental and is the fundamental property of the universe. It actually becomes all those energetic components we talked about, including the energy of the atoms coming down and the energy that created the Big Bang going out. In that nondual perspective, I see consciousness encompassing and being the source of the rest of the universe. I would consider it a combination of those three. I think that would be the best way I could explain it myself.

Kafatos: Thank you. Great. I would like to go to the next question. Lee Jae Young is next; please identify the person you are asking the question to.

Jae Young Lee: My name is Jae Young Lee, a professor at Sun Moon University. My question is for Dr. Marjorie Woollacott. My question is simple. There are many kinds of meditation in the world, but you mentioned two types: mindfulness meditation and compassion meditation. It is not empathy meditation but compassion meditation. My question is, "Which meditation do you think is more conducive for full environmental protection and improvement?"

Woollacott: Between mindfulness meditation and compassionate meditation, could one be better for environmental improvement? I would say, first of all, that when people have done research on the two types of meditation, they found that both lead to increased compassion for others. Therefore, I also believe that meditation fosters compassion for the environment, all mankind, animal-kind, and plant-kind. My own sense is that the key issue is that we are quieting the mind down, therefore experiencing our sense of interconnectedness or oneness with all the other people and beings in the universe. That interconnection actually leads to compassion because we feel they are one with us at essence. My own sense is, whatever practice of meditation one uses, as long as it brings that sense of interconnectedness, we will have a beautiful sense of responsibility for taking care of our environment, for all the future generations of the world.

Helané Wahbeh: May I add to that, Dr. Kafatos?

Kafatos: Please do.

Wahbeh: I probably agree with this, and, as a meditation teacher myself, it is a question I get quite often, such as, "What type of meditation should I do? Which one is the most effective?" My answer is that it is the one you can actually do. In other words, it is more effective if you find any type of meditation that you can actually stick with on a daily basis, even if done for just five minutes a day. We are so lucky right now, as meditation is incredibly accessible, ranging from smartphone applications to online courses. Anyone can discover a meditation practice and try another if a practice does not work for you. Whether it be mindfulness meditation, compassion meditation, focused meditation, or concentration meditation, there are so many different types of meditation that people can do to enable them to get into that transcendent, peaceful, and still-minded state. Thank you.

Kafatos: Thank you. This is actually a very important comment. There was a book that was written some time ago, called *Meditation: If You're Doing It, You're Doing It Right: Conversations with Meditators* (2015). I think that is what you are pointing to, rather than quite often what the scientific worldview would like to categorize as the tendency of the mind to say, "Oh, this is better than the other." I also think we have heard consistently from all the speakers, presenters, and commentators that this is not the case. Thank you for the additional comment that you provided. Next, we have a question from Thomas Selover.

Thomas Selover: Thank you. My question is for Dr. Wahbeh. I will try to phrase it this way. You have talked about some of the scientific evidence for

ESP or telepathy. My question is about the role of technology in terms of allowing us to research these phenomena, but also as a substitute for the phenomena themselves. For example, we use cell phones very freely now. However, according to the understanding of nonlocal knowledge or telepathy, some kind of abilities to communicate at a distance were already there before cell phones came along. For many in our culture, we are becoming so dependent on the material manifestation of those abilities, if I could put it that way. After all, the creation reflects the Creator, and we may be losing some of the natural abilities that we humans have had earlier on, before this technology. I wonder if you could comment on that.

Wahbeh: Thank you for that wonderful question. I do have a few comments on that. First, I feel that our technology has actually opened our minds to nonlocal phenomena. If you told someone 200 years ago that you could instantaneously reach someone across the world, they would not believe such a thing was even possible. Thus, I feel that this concept of instant communication, such as the Internet and use of cell phones, has actually opened the human mind for these capacities to be more easily accepted without the technology. That may be a little contrary to what you were thinking about, but I do think that these technological concepts have opened us up to it.

However, have they also blocked our natural, intuitive abilities, because they are allowing us to do what perhaps our intuition would have done in the past? I do not really know the answer to that. If you think about our technology today, it is moving so quickly, as apparent when looking at my 22-year-old son and nine-year-old son. My 22-year-old knows what a wired phone is, but my nine-year-old has no idea, having never seen a wired telephone in his entire life. I mean, just in that short time period, technology has moved so quickly.

We also have done a very fascinating study looking at the genetics of these phenomena and seeing whether certain people who are very psychic have different genetics than people who are not. It was a small pilot study, but we compared people who had a family history of psychic ability and were highly psychic to controls who did not have any family history and were not psychic. We found a noncoding region of the DNA that was in the original, wild-type form in the psychics but in a mutated form in the controls. We also looked at this very large socio-genomic database and found that there was a strong relationship between the expression of the mutation in greater technological communities. This is a very incredible thing to think about. It also aligns with your comment, which is that the greater technology, the less we were sort of staying in this wild-type form of intuition. Thus, we are definitely looking forward to researching this a lot more and seeing if we can replicate those findings. The short answer to your question is that I am not sure. I also look forward to discovering what we can about it. Thank you.

Kafatos: I would like to give the floor to Arnaud, as it seems he wants to add something. Go ahead, Arnaud.

Arnaud Delorme: Thank you, Menas. I would like to concur on what Helané said. Just to add to what Helané said, technology, for example, cameras or recording devices, has really allowed us to start studying these phenomena from a scientific perspective. Before, we had pen and paper, so it was just much harder to do so. I think it has also opened the field to the scientific validation of these phenomena.

Kafatos: If I may also say something, I think what we heard for the last two comments or answers is that technology is not the "end-all." However, it opens new vistas. It is actually very interesting what Helané said about her young children. It seems that nowadays, children are born with so many capabilities in terms of technology, such as surfing the Web and finding things online. They are doing things that my generation, who in a way were like dinosaurs, could not. So, perhaps these technologies are opening up new vistas.

As we have all heard, I think we are emphasizing that although we depend on technologies, they are actually enlarging our horizons. That would perhaps be the way that the great work IMS is doing, which really shows that these technologies can give us new ways that were not thought of even 10 years ago, I would imagine.

Let us see if there is another question. We have Takahiro Hiroi.

Takahiro Hiroi: In my impression, Dr. Wahbeh basically has pointed out scientific evidence that a person's intentions can affect someone else or even natural phenomena. You kind of suggested we could affect how we protect the environment through our intentions and mindfulness. It means that if we are not mature enough to do the right thing, we could actually harm the environment even without physically doing something.

For example, if we have spiritual power or supernatural power, it could be used for warfare or spying, as the Soviets did, as far as we know. We can develop our capabilities starting from a materialistic view, go beyond a materialistic view, and then become capable of those kinds of supernatural powers. For example, we have the Internet, and we have viruses to attack other camps, as well as even computerized weapons. How do we use our thoughts? Or, how do we think we can prevent abusing those kinds of supernatural powers in the future?

Woollacott: Those are very good questions. I will answer one of them and then go on to Helané. First of all, I really believe that when you turn down those filters on the brain, so that you feel that unity awareness with other people and with the planet, you would no longer have the desire to do these harmful things to other beings or the planet. Thus, the beauty of meditative practice and becoming more aware of your interconnectedness with the whole world is that you would not want to destroy someone who you believed was actually one with you at your own essence. I think that is the good news—the more that we have everyone experiencing that interconnectedness, the less tendency we will have to want to create computer viruses, wars, or the like. I will let Helané take it from there.

Wahbeh: Thank you for that question. I am also completely aligned with you, Marjorie, on that. What I am about to share is my personal understanding;

I have no scientific evidence to back this up whatsoever. However, I think it is mirrored in many spiritual traditions and many esoteric spiritual traditions that there is a spiritual maturity that is often there to allow these intuitive capacities to be so strong and just instantaneously manifest them in a macro way in the world. Right now, if you imagine that our filters completely disappeared tomorrow, and all of a sudden, all of your thoughts and intentions were experienced by people around you and manifested around you, the world would be in incredible chaos. Chaos would ensue because not everybody on the planet has a meditative practice, has a meta-awareness of their consciousness, their thoughts, and the like.

Yet, my hope is, as we do become more spiritually mature, as humanity matures, that our coherent, aligned intent could actually be used to support our planet and each other. As Marjorie said, from that interconnected, compassionate, loving space, there will not be the desire to make viruses or harm others because you realize that when you are harming someone else, you are actually harming yourself. Thus, the impetus to want to harm just naturally dissipates. You can call me naive or idealistic, yet I really feel that is where the future of humanity lies. My personal work and motivation is to help manifest that more quickly. Thank you.

Kafatos: Are there any further questions? I see we have one more question. We have Glenn Strait with a question. Please go ahead.

Glenn Strait: Prof. Woollacott, you have spoken extensively on unity consciousness in relation to the brain. I would like to hear your comment about the role of the heart in unity consciousness. Thank you.

Woollacott: I think that some people talk about the fact that we have to consider our nervous system as really encompassing our entire body. The fact is that we have, of course, neurons in our brain, but we also have neurons in our heart and our gut, and they are all informing us in different ways about how we are relating to the environment. I typically talk about the brain, but

I also include all those other parts of the body. I would also like to mention one other thing, which is that often in the spiritual traditions of many cultures, they talk about the heart as a subtle energy center. They mean it both as a subtle energy center in the area of the heart, which is filled with compassion. Also, I think the heart in Sanskrit is called the *hridaya*, which means the core of our consciousness, the core of the universe, and the source of all consciousness.

I think it is also the important part of us that has that sense of interconnectedness and love that is radiating from us. I believe that love is encompassed in all three: in the physical, perhaps based on the neuronal connectivity between heart, brain, and gut, as they are all working together in harmony. I also love to think of the subtle energy center, at the level of the heart, or the *hridaya*, which is the core of the universe filled with love and peace and unity awareness.

Kafatos: Thank you, Marjorie. In fact, regarding the heart and the mystical world of the subtle system, where perhaps Kundalini energy resides, one views the various energy centers as really a continuum—as something that connects everything, without a higher and a lower.

I personally feel very gratified again to have listened to these wonderful talks that were thought-provoking and heart-provoking, just to pick up on this last point. I really thank the presenters and the commentators for their great presentations. Also, I would like to thank the organizers for putting together such a thought- and heart-provoking view of the future, which you do not find very often. I will not say I am sorry, but being negative myself, such presentations are actually quite difficult to find in academia.

I also think that some of us, who are more senior people, are trying to encourage the younger generation to not be intimidated and pursue these paths that have been laid before us today, in a very clear, compassionate, and loving way. Because I think there may not be much of a future for humanity unless we do so. Actually, I am an optimist. I believe there is a way out of this morass that we find ourselves in these days. I will not make a speech fit for a pulpit, but it is really bringing us together in a community, in large communities that are connected. In any case, that is to say, we can make this science of spirituality, if I may call it that, more formal. The eternal science of the knowledge of the beyond is the way forward, and it is, as we heard, the capacity of every human being and not just the very few talented people.

With that, I think we have reached the end of our session. I would like to thank everybody again. I would particularly like to thank the organizers for inviting us all, and tomorrow I have the difficult task of presenting in 15 minutes the results of this great session. I will try to do my best tomorrow to do so at the end. Thank you everyone. ◊

ICSG III

SESSION 2

Applications of Frontier Science in Addressing Environmental Challenges

About Session 2

Applications of Frontier Science in Addressing Environmental Challenges

Scientific support for a postmaterialist paradigm comes from evidence gathered by researchers who are engaged in what is called "frontier science" or "new paradigm science." This session will examine applications of frontier science in medicine and energy production. In the course of history, various "alternative" medical treatments have been developed in different cultures of the world. Examples include energy medicine (or psychic healing), acupuncture, Ayurveda, Qigong, homeopathy, Reiki, music therapy, and aromatherapy. These treatments can help heal illnesses caused by environmental pollutants, and they usually do not have the adverse side effects of modern medicine. In this session, the first talk presents a scientific investigation of energy medicine. The second presentation reviews a wide range of innovative, energy-producing technologies based on new paradigm science–such as the use of triboelectricity (friction energy), magma geothermal (from the Earth's magma), and "vacuum energy" (or "zero-point energy"). These technologies would avoid or greatly reduce the release of greenhouse gases, help address the problem of climate change, and benefit the environment. ◊

About the Chair, Dr. Paul Filmore

- Visiting Senior Research Fellow, School of Engineering, Computing and Mathematics, University of Plymouth, UK.
- Director, The Insight Centre Ltd. (Since 2004): Specializes in systematic problem solving, including breakthrough thinking and creative thinking tools, for teams and organizations.
- Chairman of the Board, Scientific & Medical Network (Since 2015): an international educational charity and creative international forum for transformative learning and change (www.scientificandmedical.net).
- PhD in Physics, University of Surrey

Session 2 | Presentation One

The Biophysics of Energy Medicine: Effect of Polarity Therapy / Subtle Energy on Cellular and Molecular Function

Dr. Christina L. Ross

Wake Forest School of Medicine,
Winston-Salem, North Carolina, USA

Current practices in allopathic medicine measure different types of energy in the human body by using quantum field dynamics involved in nuclear medicine, radiology, and imaging diagnostics. Once diagnosed, current treatments revert to biochemistry instead of using biophysics therapies to treat the disturbances in subtle energies detected and used for diagnostics. Quantum physics teaches us there is no difference between energy and matter. All systems in the human being, from the atomic to the molecular level, are constantly in motion-creating resonance.

About the Speaker

- Board-certified Polarity Practitioner (BCPP); Registered Polarity Educator (RPE); Certified Energy Medicine Practitioner (CEMP); Owner, Polarity Plus.
- Her dissertation involved research at the Wake Forest School of Medicine's Institute for Regenerative Medicine, studying the therapeutic effects of extremely low-frequency pulsed electromagnetic fields for the treatment of pain and inflammation.
- Author of the book on Energy Medicine titled, *Etiology: How to Detect Disease in Your Energy Field Before It Manifests in Your Body* (2013).
- PhD in Energy Medicine from Akamai University in Hawaii and bachelor's degrees in both psychology and physics from the University of North Carolina.

This resonance is important to understanding how subtle energy directs and maintains health and wellness in the human being. Energy medicine (EM), whether based on human touch or device-based, is the use of known subtle energy fields to therapeutically assess and treat energetic imbalances, bringing the body's systems back to homeostasis (balance).

The future of EM depends on the ability of allopathic medicine to merge physics with biochemistry. Biophoton emissions, as well as signal transduction and cell signaling communication systems, are widely accepted in today's medicine. This technology needs to be expanded to include the existence of the human biofield (or human energy field, HEF) to better understand that disturbances in the coherence of energy patterns are indications of disease and aging. Future perspectives include understanding cellular voltage potentials and how they relate to health and wellness, understanding the overlap between the endocrine and chakra systems, and understanding how EM therapeutically enhances psychoneuroimmunology (mind–body medicine).

EM is the use of known subtle energy fields to therapeutically assess and treat energetic imbalances, bringing the body's systems (neurological, cardiovascular, respiratory, skeletal, endocrinal, emotional/psychological, and the like) back to homeostasis. Knowledge of the existence of the human energy field is the first step to understanding integral physiology, which unites body, mind, and spirit to treat the entire human being—not just the physiology. Disturbances in the coherence of energy patterns of the HEF are indications of disease and aging. When these energy particles are exposed to EM in the form of coherent energy patterns—namely, through pulsed electromagnetic field (PEMF) devices, vibrational medicine, Reiki, Polarity Therapy, acupuncture, Healing Touch, and the like—the disturbed resonant patterns in the field return to their original, coherent, harmonic, and vibrational state (homeostasis) (**Figure 1**).

If Western medicine applied the principles of modern physics, it would understand that human beings are composed of information (energy) interacting with other energy (environment) to profoundly impact our physical and emotional health. Human energy vibrations were recorded at 1000 times

The Physics of Energy Medicine

1) Waves interact with the human biofield through the principle of superposition: which states that when two different waves meet, they create a 3rd waveform.
2) As the subtle energy treatment (wave) interacts with the biofield, the net result is constructive interference.
3) Coherent field frequencies are harmonic in nature and override incoherent (noise) field frequencies that are attributed to disease (Hunt 1996).

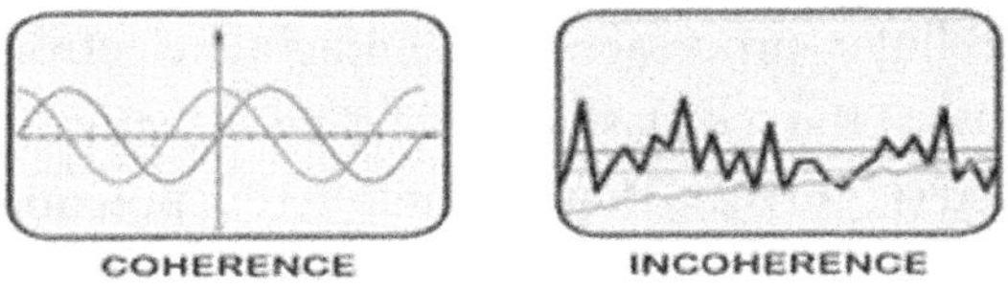

4) A balanced state of health depends on the ability to respond appropriately to physical, mental, and/or emotional stressors in the environment.
5) A balanced energy field is constantly adjusting to new stressors of internal and external environment (blood pressure, temperature, immune responses) all moving toward dynamic equilibrium (homeostasis / balance).

Wake Forest Institute for Regenerative Medicine

Figure 1

higher in frequency than the electrical signals of nerve and muscle, with continuous dynamic modulation, unlike the pulsing signals of the nervous system (Hunt 1996). Energy in the HEF is typically referred to as subtle energy. It is a system of wave-particle matter, transmitting and receiving vibrational information governing the physical matter of the body. Healing is achieved by directing coherent, harmonic energy into distortions caused by stressors and disease. For many years, Western medicine rejected the possibility that an EMF could affect biochemical mechanisms with such weak electrical fields. Biochemistry, however, is based on an understanding of the flow of energy that drives chemical reactions.

Physical properties of molecules can be combined to express internal energy and thermodynamic potentials, which are necessary for equilibrium and homeostasis in spontaneous processes (Wisneski and Anderson 2009).

New models of biophysics emphasize cooperative electrical activity of highly ordered elements at all levels of physiology: cells, tissues, organs, and organ systems, as well as the entire body. Laboratory research with in vivo (animal) and in vitro (cell and tissue cultures) has shown important healing effects by low-frequency or weak EMF therapies, causing changes in cell proliferation, with alterations in membrane structure and function. Parameters of these fields include frequency, amplitude (field strength), waveform, and time of exposure. Detailed laboratory research in biophysical stimulation has identified specific cellular processes responding to electromagnetic forces. Selective pathways at the cell plasma membrane are activated, depending on the PEMF applied (**Figure 2**). These include action potentials and activated voltage-gated ion channels (Funk and Monsees 2006; Funk et al. 2009).

According to the National Center for Complementary and Integrative Health (NCCIH, formerly known as the National Center for Complementary and Alternative Medicine (NCCAM)) at the U.S. National Institutes of Health, there are two types of energy medicine:

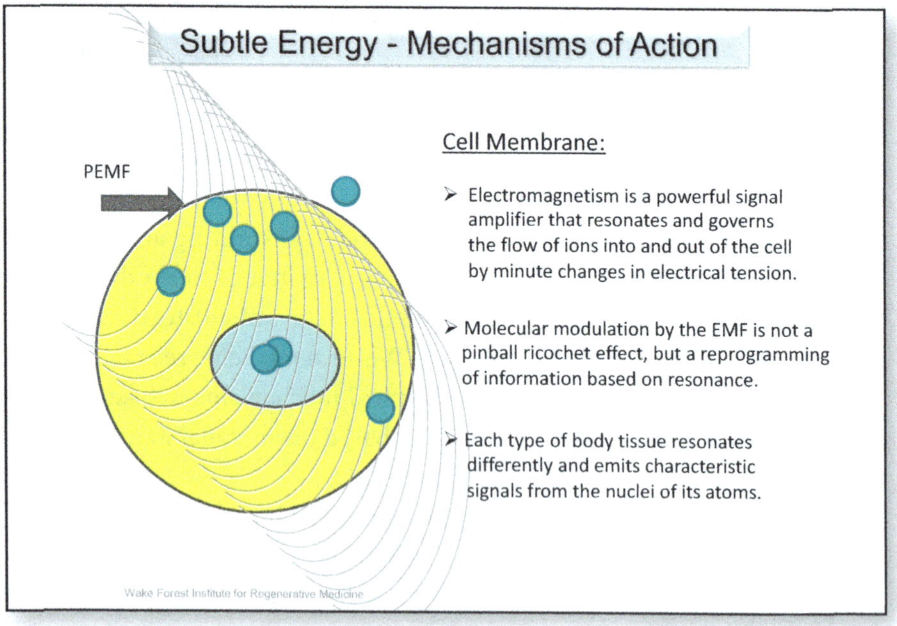

Figure 2

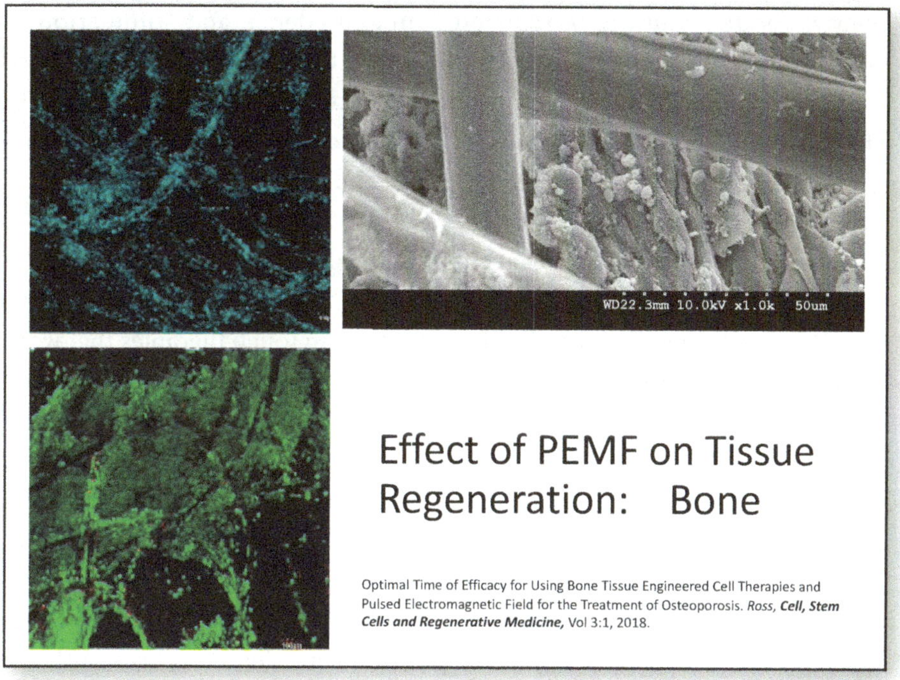

Figure 3

1) *Putative* energy medicine is based on the idea that human beings can influence subtle energy in the body using their hands, intention, and/or meditation. By focusing on these subtle energies, EM practitioners can feel vibrational frequencies with their hands and align the human biofield with healing results.

2) *Veritable* energy medicine, which uses devices that produce subtle energy fields to align the biofield with healing results.

In device-based or veritable EM, research shows PEMF at extra low frequencies (ELFs, <100 Hz) is beneficial to immune system modulation as well as tissue regeneration (**Figure 3**) (Ross and Harrison 2013; Ross et al. 2015, 2016; Ross 2017). PEMF can pass through the skin into the body's conductive tissue, resulting in reduced pain and edema, and stimulation of wound healing after trauma (Ross 2017). Electromagnetic medical device therapies can affect cell signaling systems through the modulation of cytokine function, second messengers such as cyclic adenosine monophosphate, transcription factor nuclear factor kappa B, and tissue regeneration, without cytotoxic or genotoxic effects (Ross et al. 2015, 2018; Ross 2017). EMFs oscillate at various frequencies; however, ELFs (<100 Hz) are commonly used for therapeutic purposes.

Putative forms of energy medicine are human touch therapies that work using touch, interaction, and certain protocols to modulate energy imbalances in the HEF. Polarity Therapy, Healing Touch, Reiki, Cranial-Sacral Therapy, and Trager, Bowen, and Brennan Healing Science all use similar techniques for bringing the HEF back to homeostasis. The subtle energy of the HEF is easily modulated by the therapist's hands. The client and practitioner work together using breathing techniques to move stagnant or blocked energy from the cells, across the tissues and through the organs. Human touch therapies are patient–practitioner oriented, where both the giver and receiver of the energy treatment must work in tandem for beneficial results to occur. The practitioner grounds and centers himself/herself, meaning all thoughts, emotions, and physical sensations are neutralized. During optimal healing states, our bodies resonate at certain frequencies (0.3–100 Hz), which correlate with delta, theta, alpha, and beta brain waves. Human touch therapies bring distorted frequencies of brain waves, organ resonance, and endocrine/chakra systems back into balance by modulating the subtle energies of the HEF. Practitioners detect and manipulate the subtle energy and provide a resonating template for the patient's biofield to follow. In this state, body–mind–spirit is optimal for bringing the patient/client back into balance energetically. Practitioners of human touch therapies are a valuable resource in guiding both the practice and science of biofield therapies and could, with collaborative support of researchers and medical doctors, prepare meaningful case reports and best-case series for patients (Ross et al. 2022).

The pathway from the physical body through the hormones to the psychological and emotional body is through the endocrine system, which is closely associated with the chakra system. The major glands of the endocrine system are the hypothalamus, pituitary, thyroid, parathyroid, adrenal, pineal, and reproductive organs. Overlapping the endocrine system is the chakra system, which contains seven vital energy centers that run from the base of the spine to the top of the head, centered on the spinal column. They include the following (**Figure 4**):

(1) root chakra—associated with the adrenals,

(2) the sacral chakra—associated with the generatives (ovaries for women and testes for men),

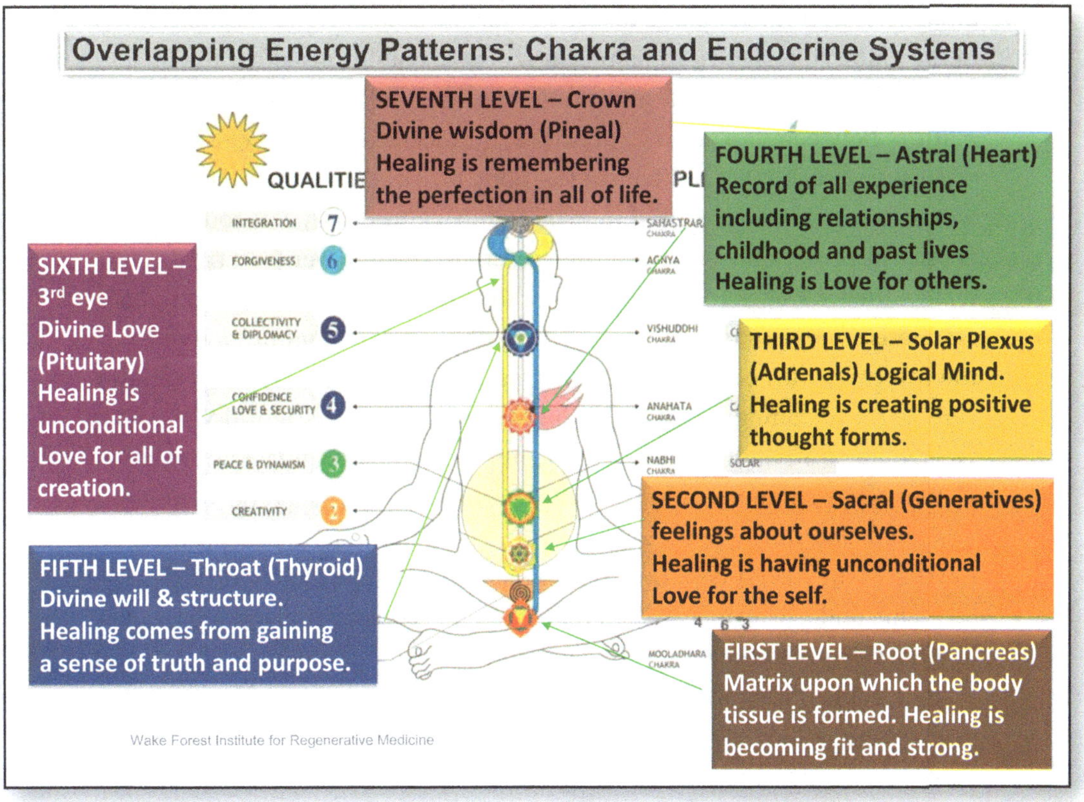

Image source: https://commons.wikimedia.org/wiki/File:Chakras_en_en_el_cuerpo_humano.jpeg ; Image Courtesy: IXSUN **Figure 4**

(3) the solar plexus chakra—associated with the pancreas,
(4) the heart chakra—associated with the thymus,
(5) the throat chakra—associated with the thyroid,
(6) the third eye (located between the eyebrows)—associated with the pituitary gland, and
(7) the crown chakra—associated with the pineal gland.

Chakras produce energy vortices, which, when healthy, provide the energetic information by which all systems of the body create a global information system. More empirical data are needed to determine whether EM therapies can heal endocrine diseases/distortions through the subtle energy in and around these glands. Endocrine disorders include glucose homeostasis disorders, thyroid disorders, calcium homeostasis disorders, metabolic bone disease, pituitary gland disorders, sex hormone disorders, and tumors of the endocrine glands, to name a few. These conditions affect the quality of life of millions of people around the world. When the exchange of information between hormones, peptides, neurotransmitters, cells, tissues, organs, and regulatory systems in the body breaks down, the most efficient way to bring this exchange of information back to health is through subtle energy information exchange. The directives of subtle energy treatments realign the HEF back to homeostasis or default mode. The HEF information system is similar to the interconnection of all global computers that form the Internet, with each cell representing an individual personal computer constantly uploading and downloading information to the World Wide Web. Once there is an understanding of the human being as a global information system of cell communication, signal transduction, and energetic instruction sets, medicine will begin to treat the entire human (body–mind–spirit) with physics as the lowest common denominator, instead of biochemistry.

One underlying mechanism that would bring biochemistry and nuclear diagnostic applications together would be an understanding of how resonance applies to biological systems. After discovering biophotons in the body, Popp et al. (1984) found that the source of biophoton emissions is deoxyribonucleic acid (DNA). They discovered that DNA sends out a large range of frequencies, where certain frequencies are linked to certain functions. They reported that biophoton emissions are low intensity because they are involved in cell coordination and communication that could occur only at quantum levels. Once energy reaches a certain threshold, molecules begin to vibrate (resonate) in unison until they reach a level of coherence. The moment molecules reach this state of coherence, they take on certain qualities of quantum mechanics, including non-locality, where they operate in tandem. This occurs in ion flux where the selectivity filter of ion channels exhibits quantum coherence, which is relevant to the selectivity and conduction of ions (Ca^{2+}, Na^{+}, K^{+}, and the like); endocrine hormone secretion, where highly organized timing of circadian rhythms and daily control of hormone secretion achieves optimal biological functioning in health; psychoneuroimmunology (PNI) and decoherence, which is associated with disease (Kriegsfeld and Silver 2006; Vazari and Plenio 2010).

PNI, or mind–body medicine, is associated with many aspects of the human endocrine system. PNI explains the connection between the mind/thought and the immune and nervous systems. Life experiences such as stressors and depression induce immunological activation, associated with cytokines and the hypothalamic-pituitary-adrenal (HPA) axis, which is the central stress response system (Suchecki 2018). PNI studies exact mechanisms through which specific brain immunity effects are achieved. There is evidence for nervous system/immune system interactions on several biological levels. The HPA axis responds to physical and mental challenges to maintain stability in part by controlling the body's cortisol level. Imbalances in the HPA axis are the cause of many stress-related illnesses. Complex interactions between cytokines, inflammation, and adaptive immune responses maintain homeostasis in the body to protect against disease. As discussed earlier, PEMF has been reported to significantly downregulate key inflammatory responses involved in neuroinflammatory diseases to provide critical results in treating dysfunction of neurotransmitters in severe stress and depression using EMF devices.

The future of EM depends on the ability of Western medicine to merge physics with biochemistry. As mentioned earlier, Western medicine uses physics to diagnose and then immediately reverts to a biochemical model to treat. Biophoton emissions as well as signal transduction and cell signaling communication systems in the body are also widely accepted in today's medicine. However, the idea of a cellular and molecular global communication system involving energy fields is beyond the central dogma of Western medicine. Future perspectives include bridging the gap between allopathic and EM. They would include the crossover between the following: (a) understanding how ions regulate cellular functions (action potentials and voltage-gated ion channels), and how PEMF can modulate these cellular functions to attain health and wellness, (b) understanding the overlap between the endocrine and chakra systems to discern how the physiological, psychological, and consciousness systems work together, and (c) understanding how EM therapeutically enhances PNI (mind–body medicine). ◊

References

Funk, R., and T. Monsees. 2006. "Effects of electromagnetic fields on cells: physiological and therapeutical approaches and molecular mechanisms of interaction." *Cells Tissues Organs* 182 (2): 59–78. https://doi.org/10.1159/000093061

———, T. Monsees, and N. Ozkucur. 2009. "Electromagnetic effects—from cell biology to medicine." *Progress in Histochemistry and Cytochemistry* 43 (4): 177–264. https://doi.org/10.1016/j.proghi.2008.07.001

Hunt, V. 1996. *Infinite Mind: Science of the Human Vibrations of Consciousness.* 2nd Edition. Malibu, CA: Malibu Publishing Company.

Kriegsfeld, L.J., and R. Silver. 2006. "The regulation of neuroendocrine function: timing is everything." *Hormones and Behavior* 49 (5): 557–574. https://doi.org/10.1016/j.yhbeh.2005.12.011

Popp, F., and S. Cohen. 2003. "Biophoton emission of the human body." *Indian Journal of Experimental Biology* 41 (5): 440–445.

———, W. Nagl, K.H. Li, W. Scholz, O. Weingärtner, and R. Wolf. 1984. "Biophoton emission: new evidence for coherence and DNA as source." *Cell Biophysics* 6: 33–52. https://doi.org/10.1007/BF02788579

Ross, C. 2013. "The use of magnetic field for the reduction of inflammation: a review of the history and therapeutic results." *Alternative Therapies in Health and Medicine* 19 (2): 47–54. PMID: 23594452

———. 2016. "Electromagnetic Field Devices and Their Effect on Nociception and Peripheral Inflammatory Pain Mechanisms." *Alternative Therapy in Health and Medicine* 22 (3): 52–64. PMID: 27228272

———. 2017a. "The use of electric, magnetic, and electromagnetic field for directed cell migration and adhesion in regenerative medicine." *Biotechnology Progress* 33 (1): 5–16. https://doi.org/10.1002/btpr.2371

———. 2017b. "Optimal Time of Efficacy for Using Bone Tissue Engineered Cell Therapies and Pulsed Electromagnetic Field [PEMF] for the Treatment of Osteoporosis." *Cell, Stem Cells and Regenerative Medicine* 3 (1). http://dx.doi.org/10.16966/2472-6990.116

———, and B. Harrison. 2013. "Effect of pulsed electromagnetic field on inflammatory pathway markers in RAW 264.7 murine macrophages." *Journal of Inflammation Research* 6: 45–51. https://doi.org/10.2147/JIR.S40269

———, T. Teli, and B.S. Harrison. 2016. "Effect of electromagnetic field on cyclic adenosine monophosphate (cAMP) in a human mu-opioid receptor cell model." *Electromagnetic Biology and Medicine* 35 (3): 206–213. https://doi.org/10.3109/15368378.2015.1043556

———, M. Siriwardane, G. Almedia-Porada, C.D. Porada, P. Brink, G.J. Christ, and B.J. Harrison. 2015. "The effect of low-frequency electromagnetic field on human bone marrow stem/progenitor cell differentiation." *Stem Cell Research* 15 (1): 96–108. https://doi.org/10.1016/j.scr.2015.04.009

———, I. Syed, T.L. Smith, B.S. Harrison. 2016. "The regenerative effects of electromagnetic field on spinal cord injury." *Electromagnetic Biology and Medicine* 36 (1): 74–87. https://doi.org/10.3109/15368378.2016.1160408

———, M.J. Pettenati, J. Procita, L. Cathey, S.K. George, and G. Almeida-Porada. 2018. "Evaluation of Cytotoxic and Genotoxic Effects of Extremely Low-frequency Electromagnetic Field on Mesenchymal Stromal Cells." *Global Advances in Health and Medicine* 18 (7): 2164956118777472. https://doi.org/10.1177/2164956118777472

———, T. Overholt, R. Xu, G. Badlani, R.J. Evans, C.A. Matthews, S.J. Walker. 2022. "Pulsed electromagnetic field (PEMF) as an adjunct therapy for pain management in interstitial cystitis/bladder pain syndrome." *International Urogynecology Journal* 33 (3): 487–491. https://doi.org/10.1007/s00192-021-04862-3

Seto, A., C. Kusaka, S. Nakazato, W.R. Huang, T. Sato, T. Hisamitsu, and C. Takeshige. 1992. "Detection of extraordinary large bio-magnetic field strength from human hand during external Qi emission." *Acupuncture & Electro-Therapeutics Research* 17 (2): 75–94. https://doi.org/10.3727/036012992816357819

Suchecki, D. 2018. "Maternal regulation of the infant's HPA axis stress response: Seymour 'Gig' Levine's legacy to neuroendocrinology." *Journal of Neuroendocrinology* 30 (7): e12610. https://doi.org/10.1111/jne.12610

Vazari, A., and M.B. Plenio. 2010. "Quantum coherence in ion channels: resonances, transport, and verification." *New Journal of Physics* 12: 1–17. https://doi.org/10.1088/1367-2630/12/8/085001

Wisneski, L., and L. Anderson. 2005. *The Scientific Basis of Integrative Medicine.* Boca Raton, FL: CRC Press, 205.

Session 2 | First Commentary
on the Presentation by Dr. Christina Ross

The Biophysics of Energy Medicine: Effect of Polarity Therapy / Subtle Energy on Cellular and Molecular Function

Dr. Hiroshi Ishii
Director, Kawasaki-Kokoro Hospital, Sendai, Japan

I am a clinical doctor working in Japan, and my specialty is geriatric psychiatry. I study the relationship between the mind and the brain from the viewpoint of Western medicine, and I diagnose and treat dementia patients. My comments are based on my analysis of the abstract of Dr. Ross's presentation, so please forgive me if they are partial or off-topic.

Dr. Ross positions modern Western medicine as allopathic medicine. Allopathic medicine is a method of treating disease with remedies, such as surgery or drugs, that produce different effects from those caused by the disease. She also points out the importance of the perspective of resonant

About the Speaker

- Director, Kawasaki Kokoro Hospital, Japan
- Specialization: Geriatric Psychiatry
- Author, *The Mind-Brain Problem and Unification Thought*
- MD, PhD in Geriatric Neuropsychology

phenomena when subtle energy acts on human health and well-being, referring to the homogeneity of matter and energy and the findings of the resonant phenomena in the field of quantum dynamics. She introduces energy medicine (EM) as using subtle energy fields to assess and treat energetic imbalance. She focuses on the human energy field to understand disease and aging. Cellular voltage potentials, endocrine and chakra systems, and the relationship between energy medicine and psychosomatic medicine are also important for the future direction of energy medicine.

I have recently learned that in the double-slit experiment using photons in quantum dynamics, it is possible to influence which side of the slit the photons pass through and to influence the shading of the waves by human thought. It has been conventionally pointed out that quantum dynamics tackles phenomena in the microscopic world that differ from the everyday causality of the macroscopic world that we are familiar with, but it may become better known in the future that the power of consciousness affects the causality of matter in the quantum world. In recent years, the development of quantum computers has been rapidly promoted and has become a national project. Quantum computers use the special properties of superposition and entanglement of quantum worlds to achieve greater power than conventional computers. Will the characteristics of the quantum world be used in the field of medicine to bring about the betterment of health and well-being to people in the future? Dr. Ross's assertion is also an attempt to introduce the cutting edge of modern physics into the field of medicine.

The history of modern medicine is dominated by mechanistic, reductionistic, and materialistic Western medicine, while traditional and folk remedies were formerly called "fringe medicine" and were regarded as heretical. However, by the end of the 1990s, they came to be collectively called complementary and alternative medicine (CAM) and have achieved some recognition in medicine with the rise of ecological awareness. Dr. Ross' energy medicine is broadly considered to be part of the CAM movement and complementary or an alternative option to mainstream medicine. Ideally, mainstream medicine and CAM should not exclude each other, but rather mutually recognize and complement each other.

Dr. Ross writes about the importance of the idea of balance in energy medicine, as opposed to the analytical and reductionistic approach of Western medicine. Oriental medicine has traditionally emphasized the need for balance for human health and well-being. In contrast to analytical and reductionistic Western medicine, Oriental medicine views human illness and health from a holistic perspective, and illness is considered as an imbalance of the whole, with the goal of treatment being to restore that balance. Oriental medicine refers to three elements: *qi*, blood, and water. *Qi* is the fundamental energy of life activity, blood is the red liquid that materially supports the body, and water is the transparent liquid that materially supports the body. Oriental people considered an imbalance of these three elements to be a state of disease. The concept of *qi* in Oriental medicine seems to have something in common with the subtle energy in energy medicine. In this sense, energy medicine should be understood not only in terms of cutting-edge physics but also in relation to ancient Eastern thought.

It is said that George Lucas, the director of the blockbuster science fiction movie series, *Star Wars,* wrote the script based on the book, *The Hero with a Thousand Faces* by Joseph Campbell, a scholar of comparative religion. In the book, it is shown that there are common patterns and events among various myths and folklore. In *Star Wars,* the characters greet each other by saying, "May the Force be with you." The "Force" here refers to something akin to the life energy that fills the universe. Those who have mastered the Force through training are called "Jedi Masters." This is just from a movie franchise, but the idea of this kind of cosmic energy has existed in many traditions in different regions all over the world. I also believe that one of the reasons why these movies have been such big hits is that people unconsciously feel a life energy or spiritual energy like the Force. In this sense, the idea of energy medicine may be said to have a certain level of universality. The challenge is how it can be used in a scientific and compelling way to help people regain health and well-being. We may need to aim to become Jedi Masters in the real environment of modern society.

On a personal note, I have been practicing mindfulness meditation for several years for my own stress control. Recently, in addition to this, I have begun to practice Reiki healing, a spiritual energy imagery meditation. I am beginning to realize that imagining the life energy that fills the universe is very important for the healing of my body and spirit. My hope is that Dr. Ross's research opens up a radically new way for people's holistic health and well-being. ◊

Session 2 | Second Commentary
on the Presentation by Dr. Christina Ross

The Biophysics of Energy Medicine: Effect of Polarity Therapy / Subtle Energy on Cellular and Molecular Function

Professor Won H. Kim
Department of Biochemistry, College of Medicine,
Yonsei University, Republic of Korea

Dr. Ross talked about Energy Medicine (EM). The balance of subtle energy is essential for maintaining human health. Disturbance of this balance in the human body can be measured and used for diagnostic purposes. However, current allopathic medicine uses only biochemical approaches for therapy.

About the Speaker

- Professor, Department of Biochemistry, College of Medicine, Yonsei University, Republic of Korea
- Chairman, Society of Applied Subtle Energy (2014–2018)
- Editorial Board Member, *Journal of Vortex Science and Technology* (since 2014)
- Current research focuses on energy medicine and water science
- Inventor of revolutionary items based on new paradigm science
- Books authored include *Science of New Paradigm* (2014) and *Water of Life: A Cure for Our Body* (2008)
- PhD in Biochemistry, Oxford University, UK (1991)

Materialism is the current paradigm in science and medicine. It presumes that we are composed of materials and that conditions of disease can be reversed only through biochemical approaches. Today's science questions and answers only what can be explained by materialism. If it cannot be explained by materialism, it is regarded as unscientific. Current science within the materialist paradigm can be compared to living in a well. Life inside a well cannot be compared to life outside, in the spacious world. As Dr. Ross suggested, the future of EM depends on the ability of allopathic medicine to merge physics with biochemistry.

I suggest that the horizons of science will be expanded if we add imaginary space to current physical space. There are many facts that cannot be explained by the current materialist paradigm. Experiments will be easily forgotten if there is no theory explaining the observed facts. If we include imaginary space, many unexplainable facts can be explained. Like quantum science in the twentieth century, we need a new paradigm for the twenty-first century. The new paradigm will be initiated by the facts that are unexplainable by the current paradigm. I propose a new scientific paradigm to explain water memory, based on imaginary space. Homeopathy, which uses water memory, has more than 200 years of history, but it cannot be explained by a materialist paradigm.

Although the late French scientist Jacques Benveniste did not know about homeopathy, he was the first to show water's memorizing ability in a scientific way. Benveniste's results sparked many investigations. The most serious of these involved double-blind research performed by four independent European laboratories in 2004. The interesting fact is that the purpose of their double-blind test was to disprove Benveniste's research, but instead they proved that his finding was in fact true.

The late Nobel laureate Luc Montagnier and his colleagues showed that DNA polymerase could recognize the wave of a specific DNA sequence of HIV transferred to water and produce new DNA copies using a technique known as PCR (polymerase chain reaction). They could amplify specific HIV DNA sequences out of pure water to which the wave of HIV DNA was transferred.

This result suggests that the wave of DNA is the physical entity recognized by DNA polymerase, even though it is not a visible (detectable) particle.

A summary of my new scientific paradigm that can explain water memory is shown in **Figure 1**. According to Paul Dirac, space is full of particles with negative energy and negative mass. My hypothesis starts from the Dirac vacuum sea, which is imaginary space. Matter consists of a physical aspect and an imaginary aspect. I named the imaginary aspect of matter as its "3D wave," as it forms a 3D field in physical space, but it is similar to the concept of *Chi* in the Orient. Oriental thinking is that every material object has its own *Chi*. Every piece of physical matter has a corresponding imaginary 3D wave with superluminal velocity, which is expressed as a field with a 3D structure in physical space. 3D fields expressed by imaginary 3D waves with superluminal velocity can be referred to as information and can be measured by biophysical approaches.

A 3D wave can be separated from matter by physical stimulation, such as vigorous shaking or pounding in homeopathy, or electrical stimulation via the Schumann wave, which is the resonant frequency of the Earth. The separated 3D wave still functions like the original substance. This is called *water memory.*

New Scientific Paradigm

Vacuum is full of particles with negative energy and negative mass, which is an area of information.
Imaginary space forms a 3D field in physical space → Information forms structure.
→ **IMAGINARY SPACE**

Matter consists of physical matter and imaginary matter (*Chi*).
Imaginary matter can be separated from physical matter by physical stimulation.
→ **HOMEOPATHY**

Separated imaginary matter forms a 3D field like that of original matter (3D WAVE).
→ **WATER MEMORY**

Every piece of matter has an imaginary part. Imaginary space does not have time and space.
All pieces of matter are interconnected with one another through imaginary space.
Life and the universe involve a dialogue between the part and the whole.
→ **HOLOGRAM**

A 3D wave can be digitized and expressed as a visual image on a 2D surface.
A digitized 3D WAVE commands imaginary space to form a 3D field of matter around a 2D image.
→ **DIGITAL 3D WAVE**

Figure 1

Water memory can also be digitized. The late Benveniste used white noise for digitization. I digitized water memory with visual imaging. A digitized 3D wave can be stored in a computer and it can be expressed in various ways. A digitized 3D wave also functions like the original substance.

A digitized 3D wave has many advantages, because it is very stable—it does not lose its function as time passes and has very wide applications. By comparison, analog-type water memory has very limited applications. But if you can digitize a 3D wave, especially with a visual image, then it can be expressed on a card, clothes, or wallpaper, so you can easily make a healing environment with printed images. A 2D image of the substance expressed on a flat surface forms a 3D field around the card (**Figure 2**).

Why is this kind of card, which is printed with a normal printer and normal ink, effective? This card commands the imaginary space to form a 3D field of the substance around the card (referred to as "UN", a name for "healing energy"). For example, a UN card that expresses P53, a tumor suppressor protein, functions as a tumor suppressor.

A digitized 3D wave can also be modulated to electricity—referred to as

UN (Healing Energy) Card

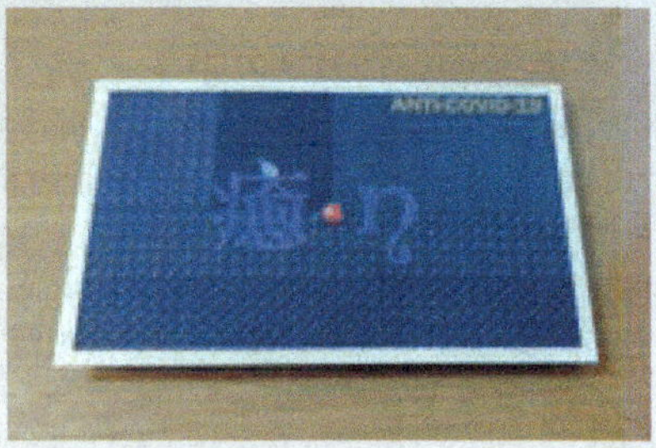

A 3D wave of matter can be digitized and stored in a computer.

UN card commands imaginary space to form a 3D field of matter around the card.

A specific 3D wave of a medically effective substance can be provided with an image (card, wallpaper, cloth, ...)

For example, when P53, a tumor suppressor, is expressed on a UN card, it functions like P53 in a human. Likewise, an anti-COVID-19 UN card protects a person from coronavirus.

Digital 3D waves of more than 1,000 substances are stored in a computer.

Figure 2

a "UL" plug, a name for "healing electricity" (**Figure 3**). By plugging the UL plug into an electrical outlet, you can eliminate harmful radiation from the Earth and purify electricity in a way that is free of electro-pollution. Specific, digital 3D waves of medically effective substances can also be put into the UL, so that the space where electricity flows will protect people against specific diseases. A specific healing environment can be provided with a variable electrical code.

The imaginary field created by the UN card and the UL plug can be used for industrial purposes and to suppress viruses. 3D waves of more than 1,000 substances are stored in a computer. They can be combined to cure specific diseases and for general health, or used for industrial purposes as well as viral diseases. The COVID-19 coronavirus can be suppressed by a specific anti-COVID-19 UN card and UL plug. The anti-COVID-19 UN card can be used for individual protection, and the anti-COVID-19 UL plug can be used for mass quarantine. New paradigm science provides a safe and effective alternative to vaccines.

The twenty-first century needs a new scientific paradigm beyond materialism to expand the horizons of science and benefit human society. ◊

UL (Healing Electricity)

Digital 3D wave can be modulated to Electricity (UL).

By plugging UL to an electrical outlet, UL eliminates harmful radiation from Earth and makes electro-pollution-free electricity that is beneficial for humans.

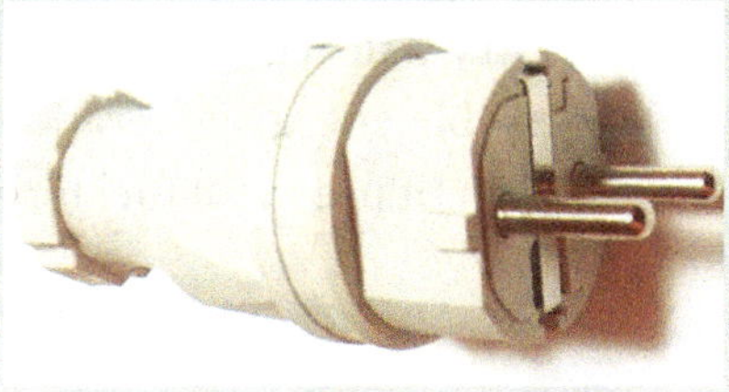

Specific 3D waves of medically effective substances can be modulated to electricity by UL.

UL with a specific 3D wave changes every electrical device to become a healing tool.

UL with P53, a tumor suppressor, makes every food item in a refrigerator an anti-cancer agent.

UL produces a virus-free environment where electricity flows.

UL with an anti-COVID-19 3D wave can be used for mass quarantine.

UL provides specific healing environments.

Figure 3

Session 2 | Presentation Two

Breakthrough Energy Technologies Derived from New Paradigm Science

Dr. Thomas F. Valone
President, Integrity Research Institute,
Beltsville, Maryland, USA

Summary

In this age of climate change, the corporate, academic, and government research into new, green sources of power and energy still has not been well-funded nor forthcoming in this century. This paper will review some of the best emerging energy technologies that promise green and carbon-free sustainable generation as well as food and water energy innovations, including a bioenergetics discovery. Progress is being made by academia and industry that will create a future having greater ease in newly developed renewable energy generation, with much better, localized energy sources that do not use fossil fuel for power and heat. Beyond the realm of fuel cells

About the Speaker

- President, Integrity Research Institute, Beltsville, Maryland, USA.
- Board Member, New Energy Movement (Since 2005).
- Inventor of numerous instruments, including PREMIER (Photonic Rejuvenation Energizing Machine & Immunizing Electrification Radiator) machine, Dental Vapor Ionizer, Powerline Gaussmeter, Static Field Gaussmeter, ELF Spectrum Analyzer, and a Geomagnetometer
- Author of many books including *The Future of Energy* (2020), *Nikola Tesla's Electricity Unplugged* (2016), *The Future of Energy* (2011), and *Zero Point Energy* (2007).
- Editor of the *Future Energy* newsletter and *Enews*.
- PhD in General Engineering, Kennedy-Western University, USA

and wind power is the unconventional world of emerging energy technologies. Some of the best examples are new and exciting generators that release trapped potential energy from nature in ways never dreamed of before. Others innovatively apply clean fuels in conventional systems that are the focus of attention for the U.S. National Aeronautics and Space Administration (NASA), Defense Advanced Research Projects Agency (DARPA), and the Department of Energy (DOE). Some of the more exotic examples include a recent breakthrough with graphene energy harvesting, which has thermal and nonthermal (quantum vacuum) sources to it.

This paper summarizes the research that has been accomplished so far in the development of highly efficient, energy harvesting inventions, as well as other future energy breakthroughs. The most exciting reason for the interest in this area of research is the promise that it holds for boosting electric vehicle production by providing an onboard electric battery charger. These energy generation developments, derived from relatively new paradigm science discoveries, also have many other applications, including a design for rural, stand-alone electrical generators. Most of them have one thing in common: they are relatively unknown to the general public.

Introduction

Current research on new energy technologies is motivated by several factors. They include:

1) The problem of global warming, which adds 1°C every 20 years due to CO_2 buildup in the air that traps heat (TinyURL.com/CO2HEAT).

2) Limited resources for food and drink, with a world population of 11 billion expected by 2100.

3) In my lifetime, since 1950 (around the time I was born), the following changes have occurred:

- The world's population has tripled (3x)
- Global carbon emissions have quadrupled (4x)
- Global energy demand or consumption has quintupled (5x)

Therefore, there is little debate among scientists and the public that we need new energy developments to save humanity.

Emphasis on Urgency: Background

Such an energy review aimed toward the future is a challenge to include remedies for power and food and waste conversion. However, at least for professionals, the primary theme of this chapter is clear and is actually the most compelling of any crisis today, since it underlies basic necessities of life, rather than a temporary threat. We, the inhabitants of planet Earth, are presently experiencing a 1.1°C increase from the 1925 baseline recorded by the U.S. National Aeronautics and Space Administration, Goddard Institute for Space Studies (NASA/GISS). It presents a moving playing field that, as of April 2020, created the second hottest March ever on record.

Carbon Dioxide Levels Drive Global Warming

In addition, the published compilation from the Vostok Ice Core (from a research station in Antarctica) of global temperature, sea level, and CO_2 levels for the past 420,000 years shows a tight correlation of all three climate variables, so that a change in one parameter affects the other two very rapidly (in geological time). No one today, except for National Geographic, is revealing the real, desperate situation that this immediate future of a six-degree rise by around 2100 presents to the world. The first graph (**Figure 1**) from the National Oceanic and Atmospheric Administration (NOAA) of global temperature will most likely become exponential with the increasing energy demand focused mainly on fossil fuel usage globally and unilaterally encouraged by the U.S. government. It is important for any environmental team to start with the premise that the CO_2 driver for this rate of temperature increase is already on an exponential increase that is unprecedented in Earth's history (290 ppm has been the maximum CO_2 for the past 400,000 years, now pushing past 410 ppm only a few decades later, with no end or "peak" in sight of the level nor the rate of change).

James Hansen provided the best compilation of atmospheric CO_2 con-

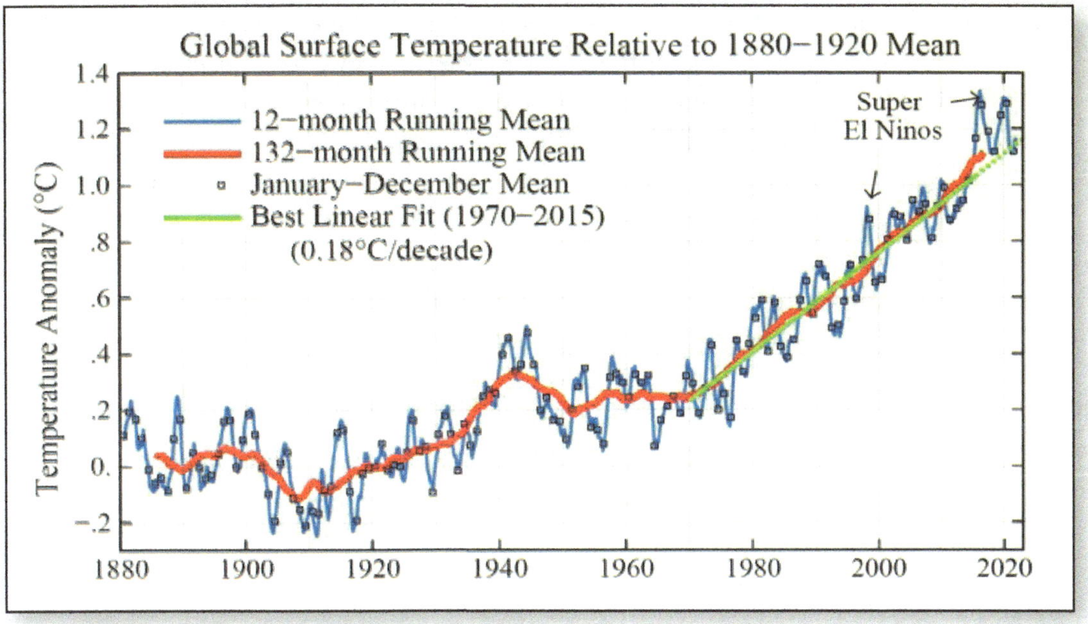

Global Warming - Manmade Temperature Anomaly **Figure 1**

Source: http://www.columbia.edu/~mhs119/Temperature/ Original data taken from NASA: GISTEMP Team. 2022. GISS Surface Temperature Analysis (GISTEMP), version 4. NASA Goddard Institute for Space Studies. Dataset accessed 2022-06-14 at https://data.giss.nasa.gov/gistemp/.
Lenssen, N., G. Schmidt, J. Hansen, M. Menne, A. Persin, R. Ruedy, and D. Zyss. 2019. "Improvements in the GISTEMP Uncertainty Model." JGR: Atmospheres 124 (12): 6307-6326, https://doi.org/10.1029/2018JD029522.

centration over time in 2006, including the off-the-chart famous "hockey stick" data point of 377 ppm in 2006, now up to past 412 ppm in 2022. The CO_2 data point in question is significantly far above the highest point in the Hansen Graph, also equal to a huge, 45% increase in worldwide carbon dioxide levels since 1850.

What we notice through the Hansen Graph is the close correlation between all three variables of atmospheric CO_2 levels, average Earth temperature, and the global sea level over the staggering period of 400,000 years. This information was gathered and analyzed by Dr. James Hansen from the Vostok Ice Core, a famous source of virgin ice from the Antarctica research station. It is also posted online conveniently labeled as www.tinyurl.com/400000years.

In determining the validity of Jim Hansen's impressive achievement, we note that it shows the distinct four past Ice Ages and the maximum of 290 ppm of CO_2 ever reached by the Earth. Furthermore, we note the clear indication of the 15°C pre-industrial global temperature maximum. Besides the thermal forcing that we humans are presenting to the world in an unprecedented short span of time, the linearity of the data correlation between the three variables is surprising, in Hansen's Table (**Figure 2**). The table actually provides the y-axis (vertical axis) grid lines for the three variables and leads one to immediately infer the simple relationship between temperature and CO_2 levels, as well as sea level (which takes centuries to materialize).

(Eq. 1): +/- 20 ppm = +/- 1°C = +/- 20m sea level

Therefore, we find that due to this author's further analysis in creating an annotated enhancement to the graph online (https://www.integrityresearchinstitute.org/CO2andClimateBeastgraph.jpg), today the present CO_2

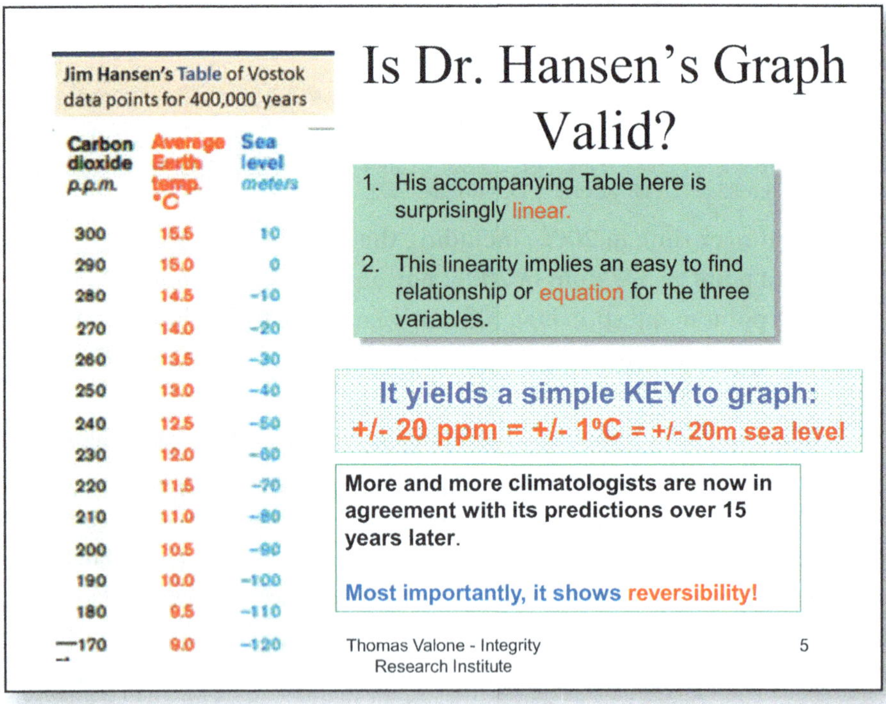

Figure 2 Is Dr. Hansen's Vostok Ice Core Analysis Graph Valid?

level of over 410 ppm minus the pre-industrial 290 ppm yields about 120 ppm of excess CO_2 in the atmosphere. We simply multiply the excess CO_2 concentration by 7.77 GT/ppm to convert this amount to gigatons of CO_2, resulting in the Earth being indebted for about 932 GT of CO_2. This must be removed by carbon capture on a gigaton level, probably over decades, to reverse the heating trend of CO_2. Thus, with each 150 GT captured and sequestered underground in a carbonate rock, we will actually realize a 1^0C decrease worldwide in our indebted temperature burden. Furthermore, by the time several degrees have been equalized and reduced by Carbon Capture and Storage (CCS), a cooling trend will start to become a welcome worldwide trend. However, realistically it will occur only a few decades from now.

One of the leading companies for gigaton carbon capture is Project Vesta, which has as its website motto, "Harnessing the power of the oceans to remove excess CO_2 from the atmosphere," with more information about their exciting technique under evaluation in the environment at this moment, on their website. Carbon-removing sand made of the mineral olivine is added to the ocean. There, the sand dissolves, countering ocean acidification and permanently removing carbon dioxide from the atmosphere. Another great company with a different approach is Carbon Engineering Ltd. in Canada. They use Direct Air Capture (DAC), which is a technology that captures carbon dioxide directly from the air with an engineered, mechanical system. DAC technology does this by pulling in atmospheric air, then through a series of chemical reactions, extracts the CO_2 from it while returning the rest of the air to the environment. This is what plants and trees do every day as they photosynthesize, except DAC technology does it much faster, with a smaller land footprint, and delivers the CO_2 in a pure, compressed form that can then be stored underground or reused.

Clean Energy Breakthroughs

A short list of the breakthrough new energy developments that have just transpired in the past decade or so include a wide range of relatively unknown sources, such as Thermal Fluctuations, Vertical Farming, Bacteria-Biomass

Electricity, Triboelectricity, Solar Lights, Water Evaporation Power, Osmotic Power, Hydrokinetic Power, Energy Harvesting, Undersea Turbines, Magma Geothermal, Graphene Fluctuating Circuits, Wireless Electricity, and Zero Point or Vacuum Energy. We will look at most of these briefly in the following sections but also provide links and references so the reader can follow up and research any particular topic more in-depth.

Looking at vertical farming, initially it can be argued that this relatively new technology offers the very best and cleanest food energy solution to the growing worldwide droughts and anticipated global food shortage, which will continue to grow as the globe keeps heating up degree by degree from manmade CO_2 pollution. Looking at the best vertical farming company in the world, AeroFarms, their most significant achievements are given by the following, with more information available on their website.

- World's largest vertical farm
- Produces 2 million pounds per year of leafy green vegetables
- Uses no soil and no pesticides
- Up to 390 times more productive than field farming
- Uses up to 95 percent less water than regular field farmers

Another solution to world hunger is the incomprehensible and unimaginable discipline of Breatharianism, using the cleanest energy source known: *prana* from the quantum vacuum. Known to be practiced in India by sadhus and saints for perhaps thousands of years, the training involved is to achieve a state of no food intake and has been shrouded in secrecy. In his book, *Autobiography of a Yogi,* Paramahansa Yogananda describes at least two breatharians, Giri Bala and Teresa Neumann, both of whom were saintly persons who stopped eating food early in life. Whether devotion to God is an essential part of the conversion from matter food input to *prana* light energy input to the body is debatable, but surely a spiritual attitude helps, from my experience. This is mainly because the process or program of becoming a breatharian automatically influences one's belief system to become less tied to material things, including food, and the depth of meditation improves. In fact, a universal experience even in the first week of training is the complete lack of a sense of hunger, which transforms a fasting experience into a *prana* development of absorbing energy from the surrounding space, both from my experience and from reports of others who have achieved a year or more of no food intake.

The book that started me taking the modern process or 21-day program seriously is *Life from Light* by Dr. Michael Werner, a biology professor, who has had trouble explaining how he is able to manufacture proteins, carbohydrates, vitamins, and the like from pure juices and other liquids like water, tea, and vegetable broth. For those interested, the book by Ray Maor, *A Year Without Food,* is the best beginner's guide. He also offers online *prana* training programs that span 7 or 9 days, with a two-week follow-up totaling the traditional 21-day program. Maor is also a good teacher and has many successful students who help give their stories during these courses. His eBook version is online for free (https://raymaor.com/ebookyearwithoutfood/). A great video and possibly the only documentary available today is, "In the beginning there was Light" on DVD and streaming at (https://www.lightdocumentary.com).

Looking toward the sun, the source of light energy input for breatharians, plants, and the Earth, there are companies now that offer small solar lights that really help to supplement and often replace the need for electricity. One of the best solar light companies, that being MPOWERD Inc., sends their "Luci EMRG" solar light to Africa with customer purchases and donations. The beauty of solar lights in places that have no electricity allows students to study at night after dark, like most of the world, so they can succeed in school. All that is required for them to do is to put the module out in the sun for about an hour before using it all evening.

Energy Harvesting

Moving on to a surprising topic of energy harvesting, we first examine the Craig Venter Institute, which tests 600 liters/day of pig waste at local farms in Escondido, CA, and Penn State University, with both generating kilowatts of electricity from human waste. BioVolt's microbial fuel cell, for example, uses GMO Geobacter strains that generate electricity, grow slowly, and do not create microbe cake. Cambrian Innovation has also tested this at the Naval

Surface Warfare in Maryland, scaling up from 2,000 liters/day to 20,000 liters/day. Therefore, it is a self-powered system that turns sewage into clean water and electricity with the help of microbes that do double duty, cleaning the water and generating biomass electricity with anaerobic digestion.

The further development of this sorely needed addition to water treatment plants is the fact that Norway has been a pioneer of industrial size installations of the sort described above. In addition, the U.S. capital, Washington, DC, also harnesses electricity from every flush with Norway's "Thermal Hydrolysis" method to convert sludge left over from sewage (with microbes) into 13 MW of electricity, which is quite respectable.

With a report published by IDTechEx, we find that the multifaceted arena of energy harvesting includes solar cells that capture CO_2 and produce burnable fuel, invented by the University of Illinois in Chicago. Potentially a game-changing solar cell, it is inexpensively converting CO_2 with sufficient voltage to split the molecule so it can be recombined into a hydrocarbon fuel. Such a solar cell splits water instead of carbon dioxide. This cell was invented by the Lawrence Berkeley National Lab, Joint Center for Artificial Photosynthesis. This double duty solar cell is called a Hybrid PhotoElectrochemical and Voltaic (HPEV) cell. It turns sunlight and water into not just one, but two types of energy: hydrogen fuel and electricity with a total of 20% efficiency. One part of the cell contributes to solar fuel generation while the other part generates electricity.

Triboelectricity (friction energy) is another unexpected market, also projected to be a money-maker in areas that involve movement of surfaces that are in contact with each other, using polymers to harvest the electrostatic energy, expected to be a $400 million market by 2027.

Atmospheric water harvesting is a growing technology for worldwide applications in areas that are expected to be hardest hit by future droughts brought on by global warming. Harvesting water from the air is a focus of Project H2E. The H2E team was a small, interdisciplinary team composed of scientists, development experts, engineers, geospatial analysts, user researchers and industrial designers who have long been passionate about the challenge of access to clean water. Drinking water access is often hindered by a range of complicated factors and structural challenges that vary from place to place, including proximity to existing water resources, local community infrastructure, governance issues and more.

Atmospheric water harvesters—devices that pull water directly from the air—have often been overlooked as a potential solution because of their relatively low outputs in comparison to other technologies, such as desalination. One of their benefits, however, is that they can be built to be small, modular, and renewably powered—decoupling them from some of the more systemic barriers to access. The team knew that such devices would never be the only solution needed, but their hope was that it could help fill a critical gap in places where safely managed drinking water was needed most. X Development LLC, the company behind Project H2E, published their findings in Lord et al. 2021 as an open access article.

Speaking of water, another area of interest for energy harvesting is undersea hydrokinetic energy, which operates around the clock with a potential of 5 TW (5,000 GW) from the world's oceans. Undersea Turbines are distributed to Wales and the Isle of Wight, and a 200 MW field of marine turbines are in the Gulf Stream. Large ocean circulation currents are the marine energy sources for hydrokinetic power plants that float above the surface of the ocean with their turbine blades below the water. Surface entry exists from the top of these buoyed turbines to all systems. They also possess a high power-to-weight ratio and low-cost deployment. Their operation and maintenance are straightforward, with a stable spar buoy vessel anchored to the seabed. Aquantis, Inc. is the leading company pursuing this clean energy technology. Other similar large-scale generators include tidal power buoy generators, ocean current electrical generators, and floating ocean wave power buoys, all of which are in production and dispersed in various ocean locations around the world.

An unexpected water source for electricity generation is evaporation. Columbia University found that energy harvested from evaporation can power much of the US, according to their study conducted in 2017. Since this type of energy is so unusual, the slideshow presentation accompanying this paper has a demonstration on a tabletop with a small car propelled only by evaporation and bacterial spores that expand when wetted. Such a tiny effect can, in bulk, produce an imbalance of the rotor tethered to the cart's wheels, thus propelling it forward. Accompanying Tenning 2015 is a video also on Vimeo (https://vimeo.com/235801232).

It may be emphasized that even the MIT Technology Review reported that an 8-cm by 8-cm water surface can produce about 2 microwatts of electricity on average, with a maximum of 60 microwatts. Thus, a large water body with an area in the scale of square kilometers can produce kilowatts of power continuously when it is not raining.

Another water-related but equally unknown clean energy source is called Osmotic Power. It utilizes ion pressure across a semipermeable membrane formed from a saltwater body, like sea water, to produce electricity. Being located near a freshwater river, Norway's Statkraft opened the world's first osmotic power plant in 2009 to produce emission-free kilowatts of electricity. An example of such an arrangement is a natural boundary created in the Alaskan Krambeck, thereby suitable for an osmotic power generation station (Feng et al. 2016). It is also encouraging that the Nanoscale Biology Lab optimized pore size to dramatically increase power output, and the Swiss EPFL reports an improved 1 MW per square meter output.

An online short video that is quite entertaining, created by EPFL, has been posted at www.tinyurl.com/SwissEPFL explaining how this technology actually functions in detail.

Creating a steamy finish to this series of water-based clean inventions is the amazing and relatively untapped source of power underground. In January, 2007, an MIT panel issued a report, "The Future of Geothermal Energy," that indicated geothermal energy as a key US energy source (Renner 2006). In the U.S., geothermal power has steadily grown, from about nothing in 1970 to about 2800 MW of electricity around 2000, supplying electricity to about 4 million people. However, the geothermal power plants, which I have seen in California, do emit large amounts of water vapor as clouds of steam. Taking this technology one step further, there is the unbelievable new paradigm science of drilling into the Earth's mantle and tapping the 10- times increased heat of molten magma. This is actually possible and has been achieved by one of the coldest countries in the world: Iceland. Iceland is the home of many volcanoes and therefore has access to molten magma about five kilometers below the surface. There, the HS Orka has been able to generate 30 MW of power, as reported in *New Scientist* (Oct. 21, 2016), instead of only 3 MW if it drilled into normal, subsurface geothermal layered rock.

Still in the category of energy harvesting is the recently discovered microscopic power generated from perpetually fluctuating sheets of atomically thin graphene. In October 2020, Paul Thibado, PhD, physicist from the University of Arkansas, announced that his team had successfully developed a circuit capable of capturing graphene's thermal (and nonthermal quantum) motion. Their circuit amazingly converted the motion into an electrical current by using two diodes in opposite directions so as to rectify thermal fluctuations. This huge breakthrough literally materializing the proverbial and elusive "free energy" was published in Thibado et al. 2020.

A few months later, in December 2020, NTS Innovations announced they had achieved a major milestone in its development of a revolutionary clean energy source: Graphene Energy Harvesting (GEH). The company is in partnership with the University of Arkansas. NTS has completed the development of its energy harvesting circuit on a silicon wafer. This technology is being packaged in a commercial chip and made available for purchase through a global network of electronics distributors. This singular achievement should be qualified for a Nobel Prize because of its unique and long lasting (perpetual) nature of its energy source at any temperature.

Graphene is 100x stronger than steel and a better conductor of electricity than copper. It is also flexible and elastic. Better still, when graphene is freestanding, it is actually in a constant state of motion. This motion is now being harvested as clean, sustainable energy. GEH is the act of harvesting energy at the nanoscale level from the naturally occurring oscillations in graphene.

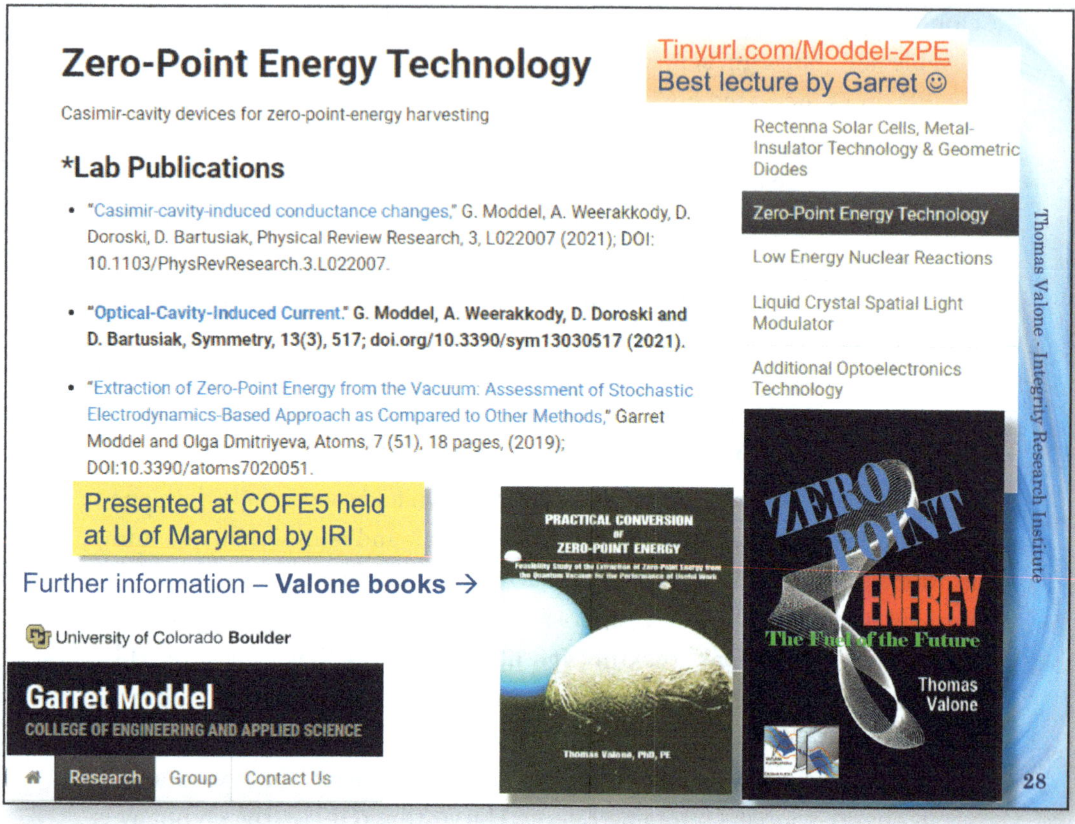

Figure 3 Various sources of zero-point energy information

To many quantum physicists, the option of seeing such a mysterious, fluctuating energy generator begs the question of whether it will also persist at very low temperatures near absolute zero. This experiment, proposed by this author to Dr. Thibado is yet to be conducted, and it would confirm a quantum vacuum zero-point energy portion of the energy phenomenon. In **Figure 3**, we see other sources of information on this new paradigm of future energy science, thanks in part to Garret Moddel, professor at the University of Colorado Boulder, who has a Zero-Point Energy Laboratory constructing Casimir cavity devices and electronic rectennas that propose to utilize zero-point energy to generate electricity.

Conclusion

In such a cursory review of a wide range of new paradigm energy generation for the future, many discoveries have been left out to maintain brevity. Many expect that in the next few years, we will be able to modify the Earth's atmosphere with Sun-shielding aerosols (that may have unexpected consequences), offering time to develop the gigaton carbon capture we need for a long-term solution to global warming. Along with the environmental solution, the development of some of these wonderful, clean energy inventions will help enormously to bring the end to fossil fuel burning and continued gigaton addition of CO_2 to the atmospheric bank above us. Further information about future energy conversion breakthroughs includes books by this author, such as:

1) *Nikola Tesla's Electricity Unplugged: Wireless Transmission of Power as the Master of Lightning Intended* (Adventures Unlimited Press), which contains the best collection of articles by the world's experts in Tesla's wireless transmission of electricity.

2) *The Future of Energy: Challenges, Perspectives, and Solutions* (Nova Science Publishers) with a great selection of future energy paradigm shifting energy discussions such as zinc oxide nanostructures, solar energy, the fission-fusion energy generator, energy storage innovations, nonlinear magnetic energy device, the Manelas electric car recharger, low energy nuclear power, a theory of time effects, and gravity/antigravity.

3) *Gravitoelectromagnetic Theories and Their Applications to Advanced Science and Technology* (Nova Science Publishers), with topics including electrogravitics, coupling of electricity and gravity, the Biefeld-Brown Effect, electromagnetic gravity control theory, electromagnetoroid structures for gravity effects, plasma and gravito-electromagnetic forces. ◊

References

Feng, J., M. Graf, K. Liu, D. Ovchinnikov, D. Dumcenco, M. Heiranian, V. Nandigana, N.R. Aluru, A. Kis, and A. Radenovic. 2016. "Single-layer MoS2 nanopores as nanopower generators." *Nature* 536: 197–200. https://doi.org/10.1038/nature18593

GISTEMP Team. 2022. GISS Surface Temperature Analysis (GISTEMP), version 4. NASA Goddard Institute for Space Studies. Dataset accessed 2022-06-14 at https://data.giss.nasa.gov/gistemp/.

Lenssen, N., G. Schmidt, J. Hansen, M. Menne, A. Persin, R. Ruedy, and D. Zyss. 2019. "Improvements in the GISTEMP Uncertainty Model." *JGR:Atmospheres* 124 (12): 6307–6326, https://doi.org/10.1029/2018JD029522.

Lord, J., A. Thomas, N. Treat, M. Forkin, R. Bain, P. Dulac, C.H. Behroozi, T. Mamutov, J. Fongheiser, N. Kobilansky, S. Washburn, C. Truesdell, C. Lee, and P.H. Schmaelzle. 2021. "Global potential for harvesting drinking water from air using solar energy." *Nature* 598: 611–617. https://doi.org/10.1038/s41586-021-03900-w

Maor, R. 2018. *A Year Without Food: Discover The Unimaginable World of Proven Energetic Nourishment.* Aingeal Rose & Ahonu, designed and eds. Prescott, AZ: Twin Flame Productions LLC.

Musha, T., M.J. Pinheiro, and T. Valone. 2013. *Gravitoelectromagnetic Theories & Their Applications to Advanced Science & Technology.* Hauppauge, NY: Nova Science Publishers, Inc.

Renner, J.L. 2006. "The Future of Geothermal Energy." United States. https://doi.org/10.2172/911903. Accessed on June 27, 2022. https://energy.mit.edu/wp-content/uploads/2006/11/MITEI-The-Future-of-Geothermal-Energy.pdf

Tenning, M. 2015. "Water, Water Everywhere." *Scientific American* 313 (3): 26. https://www.jstor.org/stable/26046224

Thibaldo, P.M., P. Kumar, S. Singh, M. Ruiz-Garcia, A. Lasanta, and L.L. Bonilla. 2020. "Fluctuation-induced current from freestanding graphene." *Physical Review E* 102 (4): 042101. https://doi.org/10.1103/PhysRevE.102.042101

Valone, T. 2016. *Nikola Tesla's Electricity Unplugged: Wireless Transmission of Power as the Master of Lightning Intended.* Kempton, IL: Adventures Unlimited Press.

———. 2020. *The Future of Energy: Challenges, Perspectives, and Solutions.* Hauppauge, NY: Nova Science Publishers, Inc.

Werner, M., and T. Stockli. 2012. *Life from Light: Is it possible to live without food? A scientist reports on his experiences.* East Sussex, UK: Clairview Books Ltd.

Yogananda, P. 2020. *Autobiography of a Yogi.* UK edition. UK: Sterling Publishers Pvt. Ltd.

Session 2 | First Commentary
on the Presentation by Dr. Thomas Valone

Breakthrough Energy Technologies Derived from New Paradigm Science

Dr. Thorsten Ludwig
Senior Scientist and Owner,
Dr. Ludwig New Energy Technologies, Pritzwalk, Germany

Thank you very much for the nice introduction and the invitation to offer a commentary on the beautiful talk by Dr. Thomas Valone on Breakthrough Energy Technologies. I, too, have been working in this field for a very long time, and I would like to give you a few more insights.

First of all, Dr. Valone is right, the planet is in trouble. If we keep going unabated, it will not be the same planet in the future. Global warming will kick in. If we keep on producing CO_2, even if we leave it at the current level, the concentration is too high. What we need to do is to change our technology. I would also like to emphasize, especially in this setting, that now we need to change our attitude towards peace, love, and kindness. Because even

About the Speaker

- Senior Scientist and Owner,
  Dr. Ludwig New Energy Technologies, Pritzwalk, Germany
- Researches and conducts experiments on the Casimir Effect, Coler Magnetic Current Apparatus, Magneto-Acoustic Resonance, and Zero Point Energy
- Doctor of Natural Science, Technical University of Berlin, Germany

if you have new technology, you can always use it for good purposes or to create something harmful. So, that is the foremost concern.

However, we also live in a sea of energy, as nature provides us with a lot of energy. We have the sun and many ways of using the energy of the sun. We have tidal waves, tidal ocean currents, and Dr. Valone showed us ways in which we can utilize these sources of energy. Thus, as shown in **Figure 1,** we can have solar energy, tidal power, and geothermal energy. Each one of the above is enough on its own to power all our machinery times 1000; so, we have a lot of energy. The top center image is the quantum vacuum, which is a kind of quantum fluctuation and a great sea of energy. The top right image refers to electricity in the atmosphere. Nikola Tesla already gave us insights on how to use that, as in the bottom right image. We thus live in a sea of energy.

Dr. Valone provided us with a myriad of interesting technologies we can use. One thing that he mentioned in the last part of his talk was the vacuum

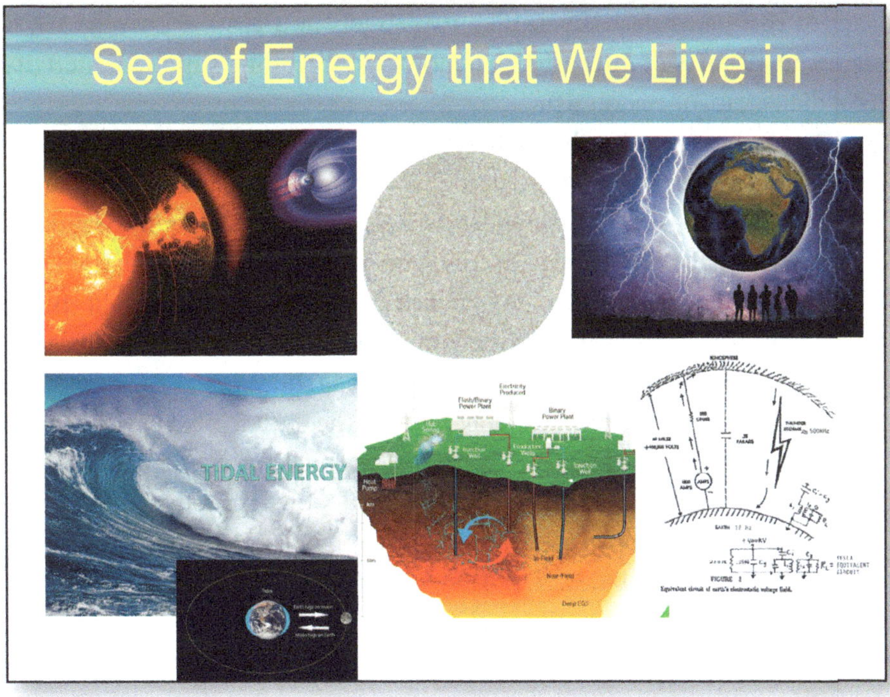

Figure 1

state. As the ancients knew it, there are five elements. The fifth element, or living energy or space energy, is a great resource. We know now that this is the Casimir effect (**Figure 2**), which I am also working on. Many people have shown that by just placing metal plates, you can have a macroscopic force due to the vacuum forces, and there are ways of using this energy.

It is also connected with magnetism. In fact, magnetism is somewhat derived from this vacuum energy. **Figure 2** shows an equation for the Casimir force, but I will refrain from delving into the mathematics in this presentation. However, Dr. Valone is totally on the right track. The far future and the near future will utilize magnetism and zero-point energy. Dr. Valone has written books on this, and that is a great thing.

Figure 3 shows a German invention that Dr. Valone did not mention. It is the Coler magnetic current apparatus, which also uses this kind of energy.

Another modern technology from the U.S. and from Germany is graphene technologies. Graphene is a 2D quantum material, like a single layer of carbon atoms. If you take two sheets and twist them slightly, it results in all sorts of interesting phenomena, including superconductivity, total dielectric, or magnetism.

When there are two sheets, they vibrate due to different environmental energies. So, the keyword here is energy harvesting, which means collecting the energy around us. With these graphene sheets, it can be done quite beautifully.

Dr. Valone also emphasized the research of Prof. Thibado, at the University of Arkansas, who showed very surely and clearly that these graphene sheets can be used to harness energy from the environment and power small devices.

That is the future of all this. As Dr. Valone also showed, you can have small devices with the central power supply, without the need to plug in your phone. It has a small sustaining battery together with a small power source. The same holds true for electric cars: you can have a small power supply in the car that uses the space of the car to collect the energy you need. This is the future.

Eventually, we will be able to create airplanes and small flying vehicles for adults and children. In the future, within a beautiful land of peace, we can fly in these small, personal vehicles powered by new energy technologies.

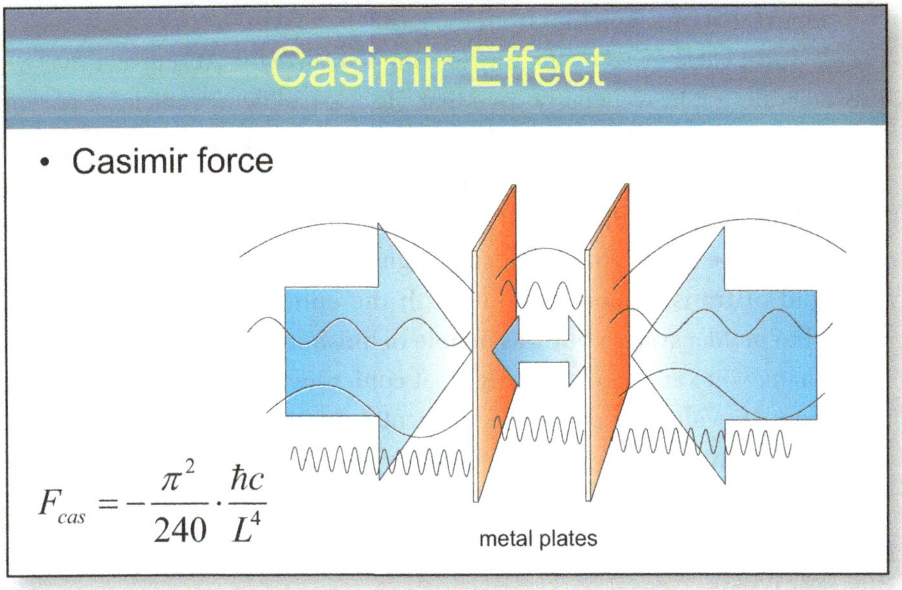

Figure 2

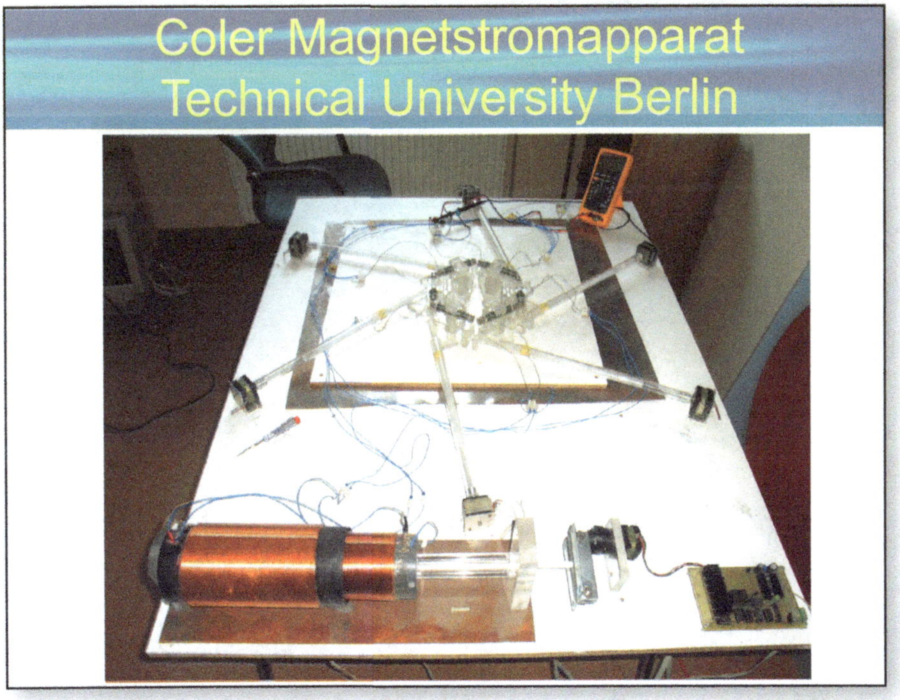

Figure 3

Figure 4 shows a picture from the past. One of the heroes in this field is Nikola Tesla. He showed us the way to commercialize the AC current that we use today. In fact, he went far beyond that. He also had flying vehicles as part of his vision. They are different from airplanes, however, due to the use of a different mechanism of propulsion, with the key being wireless energy transmission.

Nikola Tesla also discovered that, in certain ways, electricity can be collected from the atmosphere if electromagnetic waves are sent through it. It is a kind of transmission together with the enhancement of power. He started to build a system but was unable to finish it for several reasons.

Thank you very much for this great conference. I hope you all learned a lot from Dr. Valone and from my presentation. We can have a wonderful future, but remember, we need to change our attitude. We need to really want peace, live in peace, and be kind and friendly. While technology will not change our attitude, technology will help us if we truly change. Thank you very much. ◊

Figure 4

Session 2 | Second Commentary on the Presentation by Dr. Thomas Valone

Breakthrough Energy Technologies Derived from New Paradigm Science

Dr. Patrick G. Bailey
President, Institute for New Energy, California, USA

Dr. Valone is correct in all the points in his paper. We need to clearly understand all his information and his conclusions—especially about the effects of increasing CO_2 on global warming.

God, Nature, and Free Energy Devices

God, by definition, is the driving force on our Earth and us. Although we cannot see God, we can see the effects of God in our world. What we see is what we call nature.

We study nature to understand how things work, and we make laws of mathematical equations to describe what we can see. We can use these math-

About the Speaker

- President, Institute for New Energy, California, USA
- ENNEAD Executive Partner, CFO
- Senior Member of the Advanced Power Systems Technical Committee, American Institute of Aeronautics and Astronautics (AIAA)
- PhD, Nuclear Engineering and Physics, Massachusetts Institute of Technology, USA

ematical models to predict what we see and could see—within our limited understanding—inside of our "box." (Here is an example. Draw 9 dots in 3 rows of 3 each, with the dots spaced equidistant from each other. Draw a square box around the dots. Now connect all of the dots using 4 straight lines without lifting the pen.)

We should not use such models to say that we cannot expand our understanding! New surprises are found when we do research "outside the box." Examples of "impossible" inventions are steam-powered cars, airplanes, and nuclear reactors.

A "Free Energy (FE) Machine" or an "Over Unity (OU) Device" is a machine that appears to produce more energy than it uses to run. Of course, energy cannot be created or destroyed—it can only be converted from one form to another, such as from potential energy to energy used for work. Examples of such devices are dams, explosives, and nuclear reactors, which convert God's given potential energy into useful energy. For any new FE or OU device, you would not understand how it works because it is new.

There is now a real need for such new devices to solve the growing problems of global warming, the decline of bee populations, and the loss of drinkable water!

Research and Conferences, Past and Present

The Integrity Research Institute is an example where much research has been conducted on FE and OU devices. Other groups have also been researching these fields, such as the International Symposium on New Energy in the 1990s, and the Institute for New Energy (INE) since 1994 to date. The INE website was frozen in December 2003 due to U.S. Patriot Acts I and II, to ensure that it would not convey any "new technical information to the enemy." The INE website contains a lot of information from conferences, past papers, and a review of a lot of FE and OU devices that were being studied at that time, sorted by their names and commercial status. It is still up at http://www.padrak.com/ine/. As these topics are considered "fringe topics," funding is hard to find, but research is nonetheless continuing.

Research into these topics is presently being done by people in academia and industry. The research is open and unclassified, up to a point. If the research is done under a U.S. government or agency contract, the project and the results can become quickly classified. This has happened to a number of researchers in the academic, university, national, and even private laboratories. A good example is the Beta Cell Battery developed by Paul Brown, and presented at the IECEC in 1991. (His paper is on the INE website.) He showed that a resonant circuit could take the electrons from a beta-decay isotope (Sr-90, in this case) and convert the electrons to work in a small "nuclear battery." His research was shut down, he was arrested on false charges, and he was placed in jail for two years. His invention is now being worked on at Sandia Laboratories. Even with minor setbacks, progress is still being made.

However, there are still major problems. Such research is not being suppressed by the government. Rather, it is being heavily suppressed by the National Defense Networks. It is a known fact that when a patent application is submitted to the U.S. Patent Office, it is reviewed by a group of military staff, and if anyone thinks that that method or device under that application can be used to give the U.S. an "advantage over the enemy," then that patent is immediately classified! On the one hand, this is good and stops the spread of such technology to "the enemy." On the other hand, that technology is never made public—and the price of gas is still going up! There are thousands of such patents that have been classified, and the inventors are told to send in all their material and to never discuss their technology under pain of fines and imprisonment.

Funded Black Programs

The problem with such advanced technologies that are classified is that they never or very rarely get released to the general public. They are held in "Black Programs." They usually sit in a safe somewhere, and military personnel rotate into the office and then leave in a few years to a new assignment. The technologies get buried. Is this for the good of the nation, or for the future of the nation or the world? If some of this technology can be used to solve the now-pressing problems of global warming and finding freshwater supplies, then it should be carefully released in a win-win corporate business model immediately.

There are many interesting FE and OU devices that have been shown—or are reported to have been shown—to actually work. Even if they are small, they demonstrate a "New Principle of Nature" that could be expanded into working power plants for households or communities. A number of them are listed and cross-referenced by name and by commercialization criteria at the INE website (www.padrak.com/ine/db/DEVICES.html), where each device has its own data page, with all the data available for that device at that time in 2003. In addition to those devices, here are a few new devices and subject areas that are being researched today:

- Bessler's Wheel Using Gravity (1717) (Now being computer simulated to find specific parameter solutions for operation)
- Tartaria (Country) Earth Energy Electrostatic Energy Grid (mid 1800s) (Use of electrostatic energy for heating and propulsion)
- Hans Coler Current Generator (Wire-wound magnets, with current flowing into six magnets in a hexagonal design)

Hope for the Future—How We Can Make It Happen

Before now, we could rely on the world's supply of oil, and the backing of the U.S. dollar through the cost of oil, known as "Petro Dollars." As long as oil flowed into the world and into the U.S. economy, the dollar was supported and the economy was strong. Therefore, anything that threatened that flow and the use of oil in the U.S. was suppressed—taken away and classified. However, the rising amount of CO_2 caused primarily by this continuous burning of oil is increasingly attacking our environment.

If this increasing CO_2 trend continues, it could threaten our way of life, our natural environment, and probably all the Earth's various civilizations. We need water, food, and energy, excluding energy that results in continued and increasing amounts of CO_2 emissions into the atmosphere.

We need to discover new forms of potential energy and develop the proper transducers to convert that existing potential energy into usable energy. The experiments and devices noted above (and listed on the INE website) are good starting points. We also need to declassify existing U.S. patent applications and get those technologies, numbering in the thousands, into the public sector. Now some of these devices have been found to lose weight when they operate. One rule (at least) will be that any antigravity devices or effects will not be released into the public sector. Some of the devices themselves may also be too dangerous to be released into the public sector.

A simple solution is to create research teams to scan these black projects and apply them to making usable energy for our public energy grids. Corporations can be formed to build these devices away from the public sector and place them inside existing nuclear plant security fences where they can be isolated and protected. This is a win-win situation for the military, power companies, for the nations of the world, for global warming, and perhaps for our very way of life—perhaps for our own future existence.

Ask For It, Seek These Solutions, Knock On Doors!

We can all work together and make this happen. U.S. President Ronald Reagan once stated, "I occasionally think how quickly our differences worldwide would vanish if we were facing an alien threat from outside this world." However, we do not need "an alien threat"! We have made our own threat, and it exists from above: the increasing CO_2 levels in our atmosphere. We had better recognize and stop this threat immediately! All informed scientists and the United Nations agree. So, what shall we do?

Unfortunately, this is not a movie. It is not going to go away. God and Nature would say, "This is real—happening to you today! Now you know the problem, the consequences, and the possible solutions. Fix it! Help us fix it!"

Thank you for your attention. ◊

Reference

A four-page version of this Commentary—with more devices listed—can be found in English, Korean, and Japanese here: http://www.padrak.com/tvcommentary2022/

Session 2
Discussion

Paul Filmore (Session Chair): We now enter the question-and-answer session. First, I would like to offer my thanks to Dr. Ross for taking on the health and well-being part of the theme, and to Dr. Valone for taking on the theme in terms of energy, food, and water. Of course, what we are talking about is security in these domains, and what we are facing is fear. But that fear leads to opportunity, as our last commentator said. Unfortunately, fear inhibits our breakthrough thinking and stops us from also exploring new scientific paradigms. That is why moving into this postmaterialist era is very important. I agree with the last speaker. But I would also like to say that I think large corporations also have a responsibility because they sit on patents for their shareholders' profits.

I remember seeing, many years ago, that suddenly, Sanyo bought out a washing machine that was based on ultrasonics, which is a very cheap and efficient way of removing dirt off clothes. If you use ionized water to remove the dirt, there is no need for gallons of water churning through the system through multiple rinses just to remove the soap. Therefore, there is no need for the large energy level required by a spin, and so on, through all those rinses. There are many patents around that will help address our present situation. I would like to mention one other organization, as we are constantly talking about new scientific paradigms and how the worldview needs to move to this new, postmaterialist era. Pertaining to this, one organization that puts on a couple of webinars every week on this direct subject area of this conference is called the Scientific and Medical Network.

Perhaps we can post the URL into the chat. That is important because we need the general public to look in and investigate these areas and think about what this postmaterialist era will be like. So, let us move on to some questions. Again, thank you to all the presenters for their absolutely excellent presentations. Kim Jin Choon. I apologize if my pronunciation is incorrect.

Jin Choon Kim: Okay, thank you very much. I would like to ask a question to Dr. Ross. According to conventional medicine, usually we say "mind medicine" and "body medicine." In body medicine, usually in view of physics, we need to consider electromagnetic fields, as well as chemistry, biology, and medical science. However, energy medicine is somehow different from conventional medicine. In your presentation, the energy is a separate energy, a very weak and subtle energy, which brings me to my question. What is the identity of that subtle energy and its origin?

Does it have some relationship with de Broglie's theory and David Bohm's theory of the pilot wave? Or is this something different from that? Of course, it is not material energy, as it is different from conventional energy. I would thus like to know the difference between your energy and conventional energy in body medicine, and if it has any relationship with de Broglie-Bohm pilot wave.

Filmore: Thank you very much. Over to you Dr. Ross.

Christina Ross: Yes, thank you for your question. It is an excellent question. When I began studying energy medicine, I learned from very ancient traditions of medicine, including a combination of traditional Chinese medicine, Ayurveda medicine, shamanism, and ancient Egyptian medicine. And these are time-tested medicines that have been known for tens of thousands of years; indigenous people have known for tens of thousands of years that human beings have energy fields. Every one of these modalities that we studied, whether it was hermeticism, Ayurveda, or the like, said the same thing in that this subtle energy comes from the source. Now you can call this source God, the Prime Creator, or the First Cause, but it is in everything.

It permeates not just human beings but everything that exists in the universe and beyond. Some traditions call it *chi*, some call it *Qi* or *prana*. Whatever you want to call it, the subtle energy exists in and is the basis for everything. This physics drives everything: it drives the mental, emotional, spirit, consciousness, and physical body. From what I can tell from my work with subtle energy, the cells of the body respond to it. The cells of your body respond to your thoughts and consciousness, and you can direct matter with your consciousness. Avatars from the past have done this, whether you are talking about Christ, or Buddha, or whomever—they understood the dynamics of physics very similar to the way Tesla did.

You can change matter by changing its density. We need to bring back these ancient traditions into our medicine because they have been time-tested and they work—if they did not work, we would have gotten rid of them. We are missing a huge piece of the human organism, the human being. To answer your question, yes, the de Broglie wavelength phase relationships all play a huge part of it. I think we perhaps need to find a community or communal verbiage, where we are all speaking the same language about the same substance. I think that is where we kind of fall apart because we are talking about the same source of the same thing, but we call it by different names. It would really help all over the world, for medicine, to see the spirituality and consciousness in the human being in order to heal the human being. I hope that answers your question.

Filmore: Thank you, Christina. It was interesting, a little bit earlier, when Dr. Ishii from Japan was talking about *Star Wars,* and how through the phrase, "May the Force be with you," George Lucas was actually passing that concept into the general consciousness. But you are right. Our real challenge is mental models, and then changing people's mental models to understand the Western and Eastern modes of thinking and bringing them together. I see another question, from Didier Guignard. Please say who the question is addressed to.

Didier Guignard: Yes, my question is also to Dr. Ross. My question is whether she knows about the works of Dr. Bruce Lipton, and her thoughts about it; as well as a question to Dr. Ishii, who is a specialist about the mind and brain, about the work of Sir John Eccles on the relationship between conscience and our brain. Thank you.

Ross: Another excellent question. Yes, I have met Bruce Lipton, and he is a character. He and I have talked about how consciousness does affect cellular structure. You know, he started out as a cell biologist, and he was basically run out of a medical school faculty because he kept insisting that genetics can be affected by environmental factors and conscious thought. And now it is common knowledge that genes are not carved in stone and that we can affect them. In fact, I work with a lot of geneticists, and they are the first to admit that genetic imprints or genetic outcomes are just a snapshot in time, and that our genes are all always responding to what is going on in our environment. It is also a very interesting concept that all mammals on planet Earth have 24 chromosomes, while we only have 23. Right?

This is a topic for discussion because Dr. Lipton says that there is sort of a twenty-fourth chromosome, but it is the template or driver of the rest of your genetic makeup, based upon how you relate to your environment. Are you afraid of it? Are you a victim? Do you feel empowered? Do you feel like you can be the director of your own experience? These are thus the things that need to be taught in medicine. I mean, these are not just spiritual concepts. These are mental emotional concepts that people can really understand—you are not a victim, and you can create whatever dynamic you wish. As for Dr. Eccles, I am not as familiar with but am interested in his work.

Filmore: Let us move on. Next, we have another question from Takashi Kato.

Takashi Kato: I have a question for Dr. Ross. Thank you very much for your very interesting lecture. I am Takashi Kato from Japan. As I understand it, there is a form of energy, at least on a global scale. You talked about the chance to apply it to human healing. There is a flow of energy, and we can achieve human healing by changing it in various ways. We can work on the largest grade, for example, by working on the flow of energy on the global or cosmic scale. It may be possible to change our state, as our physical state is the state of the entire Earth and universe. Has there been any research or attempt in this direction? That is my question.

Ross: I apologize. I got a little bit lost. Can you repeat the last part? I heard about cosmic energy, and that is where I got lost.

Kato: You change the flow of energy to heal humanity. So, theoretically, can we change the flow of energy on a global or cosmic scale, in order to make the state of ourselves or the universe better?

Ross: This is a great question. There are a couple of physics concepts that are involved here, one being "entanglement." Entanglement has to do with everything in the universe being connected to every other thing in the universe. One powerful way to do it is through mass consciousness. There is a researcher named Rollin McCraty, who runs HeartMath. HeartMath is taking many measurements on how people who put their consciousness and minds together can influence the entire globe. This is a very powerful concept that I think more people would benefit from his work. He takes different measurements based on the energy in places that he has set up, with stations all over the planet.

Without getting too far into it, knowing that *Chi*, the source energy of the universe, is within all of us, we can use the power of this energy to benefit the entire planet. That is why, when group meditations take place for peace on Earth or for healing in certain places, there are powerful outcomes. I would invite people who are interested in changing energy all over the planet, to meditate in groups. That amplifies our power to change the consciousness of the planet. I hope that answers your question.

Filmore: Thank you. Let us move on and get some more questions. Hiroko Oizumi? Could you say who your question is for, please, and keep it brief? Thank you.

Hiroko Oizumi: Yes. Thank you very much for the excellent presentations and commentaries. My question is related to Topic One. Japan, Korea, and China have traditional medicine in common, which originated in China maybe more than 1000 years ago. However, in present-day China and Korea, they have two types of medical doctors. One is medical doctors trained in Western medicine, with the other one being doctors in traditional medicine. However, in Japan, about 140 years ago, the modernized government chose only medical doctors trained in Western medicine. I would say that, as a result, traditional medicine has been relegated to second-class status.

This is just my own experience, about a decade ago, from when I was a member of the Japanese parliament and in the Democratic Party.

We were working on medical and social insurance, which was in the red. We asked Tsumura-juntendo, the biggest company which produces traditional medicine products to submit scientific results on traditional medicine. To a certain extent, they did have scientific materials, but they did not cover everything. In the meantime, we found that many people rely upon traditional medicine, so we gave up the idea of not applying traditional medicine to social insurance. This is just my experience. I think Dr. Ross already answered this question to a certain extent from the first question, but may I repeat the question? Can the theory of energy medicine explain any other types of traditional medicine? I hope Dr. Ishii would comment on this issue as well, because he talked about Japanese oriental traditional medicine. Thank you so much.

Ross: Yes, thank you for your question. I heard at one point from a traditional Chinese medicine doctor that traditional Chinese medicine doctors are paid if the patient stays well but do not get paid if the patient is sick. I love this concept because it focuses the energy on health and wellness, instead of giving the energy to the disease. If you are fighting the disease, you are giving energy to the disease. I love this concept. Now, in polarity therapy, we study the Ayurvedic elements, those being ether, air, fire, water, and earth; and I know that in traditional Chinese medicine, they study earth, water, fire, metal, and wood. But they are the same concepts—these concepts are part of the planet. These are the elements that drive the very globe we live on. If the global elements are out of balance, you get floods, fires, and all the problems that we are having.

If our body's elements are out of balance, you get sickness, pain, and disease. I mean, we are echoing each other. It is sad that modern medicine, first of all, does not see that, and that it is not willing to embrace both modern medicine and indigenous medicine. We have an innate knowledge of who we are in body, mind, and spirit. I understand why biochemistry took over. It is because we started taking our bodies apart and looking at the organs, cells, chemicals, and molecules; it makes sense. However, we need to move past that. We need to move past medicine that can make us sick, with side effects. Energy medicine and traditional medicines balance the body. That is why the side effects are minimal, if any at all. To your point, let us not fight with each other, but instead let us join forces and get the best of both worlds. That will create healing not just with humans but with the planet.

Filmore: Thank you, Dr. Ross. That is a super-great point on which to finish our session. It is such an important area. I want to thank the organizers of this conference for enabling us to ask these types of questions. I hope that these questions will be answered in the very near future. I am sure that there will be a need for many more conferences in this subject area in the future. Again, I would like to thank Dr. Christina Ross and the two commentators, Dr. Hiroshi Ishii and Dr. Won H. Kim, as well as Dr. Thomas Valone, and the two commentators Dr. Thorsten Ludwig and Dr. Patrick Bailey for the second topic. Thank you, everybody. And thank you also to all the organizers behind this conference. It was absolutely superb. These are the questions that need to be asked, and we need to work on. Thank you again. ◊

ICSG III

SESSION 3

Frontier Science Perspectives on Our Relationship with the Natural World

About Session 3

Frontier Science Perspectives on Our Relationship with the Natural World

How does frontier science characterize the relationship between humans and the natural world? Conventional science, based on materialist thought, conceptualizes humans and all things in the world as individual entities separated by space. On the other hand, frontier science is based on a postmaterialist paradigm, which sees such separateness as superficial. It maintains that, on deeper, more fundamental levels, humans and all things in nature are interconnected. This would imply that we are also interdependent. In this session, the first talk points out this connectedness, touching on not only the external, ecological level but also the deeper, spiritual dimension. Referring to the latest concepts in quantum physics and ancient spiritual teachings, it points out the need for us to change the ways we treat and utilize natural resources, for the benefit of humanity and the entire planet. The second presentation shows us how the postmaterialist paradigm can help us live in harmony with nature. If we treat nature as sacred and express divine love toward all created things, then the Earth's future will be very bright. ◊

About the Chair, Professor GunWoong Bahng

- Leading Professor, Department of Mechanical Engineering, The State University of New York, Korea (SUNY Korea) (2014 - present)
- Principal research scientist, KRISS (Korea Research Institute of Standards and Science) (1989-2013)
- Director, National Center for Standard Reference Data, KRISS (2009-2011)
- Director, Office of Global Partnership, KRISS (2006-2008)
- Author of two books (*New Science Changes the World, Ki Moves the World*), coauthor of eight books, translator of four books
- Currently studying water structure and searching for new materials that enhance life energy in people.
- PhD in Materials Science and Engineering, Northwestern University

Session 3 | Presentation One

Exploring the Connections between Humans and Nature

Professor Chris Laszlo
Professor of Organizational Behavior, School of Management, Case Western Reserve University, Cleveland, Ohio, USA

Welcome to this presentation on "Exploring the Connections between Humans and Nature." A more specific title would be, "Quantum Physics and Associated Connections between Humans and Nature, and Their Implications for Managers."

There is solid scientific evidence showing the benefits of spending time in nature. Of course, we also know intuitively that a beautiful day outside, among trees, birds, flowers, and perhaps gently flowing water, can increase our happiness and sense of well-being. Let us begin with only a few minutes of being present to nature with its sights and sounds (**Figure 1**).

I hope that, although we were not physically in nature, these few

About the Speaker

- Professor, Organizational Behavior, Weatherhead School of Management, Case Western Reserve University, Cleveland, Ohio, USA; He researches and teaches flourishing enterprise
- Elected a "Top 100 Thought Leader in Trustworthy Business Behavior" by Trust Across America™ in 2012
- Fellow of the International Academy of Management
- Current Chair of the Management, Spirituality, and Religion (MSR) Interest Group, Academy of Management (2021-2022)
- Co-founder, Sustainable Value Partners, USA
- Author of *Quantum Leadership* (2019), *Flourishing Enterprise* (2014), *Embedded Sustainability* (2011), and *Sustainable Value* (2008)
- PhD in Economics and Management Science, Université Paris Cité

Figure 1 **Video link:** https://www.youtube.com/watch?v=5lCRsLjMeso&ab_channel=TopRelaxMusic

minutes of being present to these sights and sounds of nature already bring some benefits to you. How do we actually engage in such a connection with nature? There are some easy steps for nature immersion, as listed in **Figure 2** (Wild Ozark 2017).

Walk until you find a good place and then just listen to the natural sounds around you. Feel what is going on around you. Look up, and perhaps you can start with a tree or living movement that caught your eye, and just "be." I want to share this beautiful thought from John Muir, a well-known American naturalist, in his book, *The Mountains of California*. "Nature's peace will flow into you as sunshine flows into trees. The winds will blow their own freshness into you, and the storms, their energy, while cares will drop off like autumn leaves."

How much time do you actually need in nature to notice a difference?

Figure 2

According to one study, five hours per month is sufficient to improve mood, vitality, and feelings of relaxation (Williams 2017). This is equivalent to 15 minutes a day for five days per week.

The good news is that it can be in a city park, just as easily as in a forest or grassy plain. Even being present in a park with paved paths and street noise will show benefits. Why is this so important? One recent survey (The National Trust 2016) found that almost 50% of preschoolers lacked even one outdoor play session per day. On average, children aged 10 to 16 in the United Kingdom now spend only 13 minutes a day on vigorous outdoor activity, compared to 10 waking hours relatively motionless. This was a UK study, and I think it covered students in both the United Kingdom and the United States. So, after this long period of time of living in nature, in a very short historical period of time, we have gone to largely indoor living.

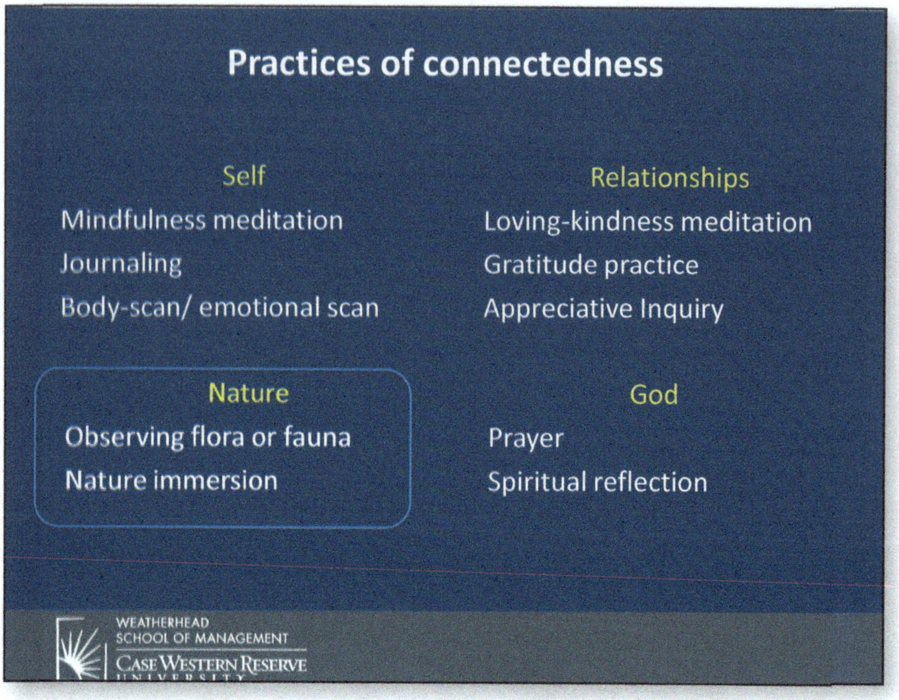

Figure 3

Harvard naturalist E.O. Wilson is one of many scientists to put forward the hypothesis that nature has a restorative power (Wilson 2009). He noted that we have a natural affiliation with nature that is ingrained in our biological heritage. Phrased more poetically, the pioneering environmentalist Rachel Carson wrote (Carson, 1962):

> *Those who contemplate the beauty of the Earth find reserves of strength that will endure as long as life lasts.... There is something infinitely healing in the repeated refrains of nature—the assurance that dawn comes after night, and spring after winter.*

I would like to point out that nature immersion, or connecting to nature, is one of many practices of connectedness. Such practices are mentioned in

Figure 3. There are practices that connect us to ourselves, such as mindfulness meditation, or journaling, or doing a body-scan or emotional scan. There are relationship-type practices of connectedness, such as loving-kindness meditation, gratitude practice, and appreciative inquiry. Also, there are practices that connect us to God or the transcendent, which include prayer and spiritual reflection. Then there is nature.

In the Seiwa Prefectural Forests of Japan, city residents practice *shinrin-yoku* or "forest bathing," which means taking in the forest air on long walks. Forest bathing became popular in Japan in the 1980s and today is a recognized preventive health-care practice. One research study measuring the physiological effects of forest bathing on 280 subjects in their early twenties concluded: "Forest environments promote lower concentrations of cortisol (a hormone associated with stress), lower pulse rate, lower blood pressure, greater parasympathetic nerve activity, and lower sympathetic nerve activity than do city environments" (Park et al. 2010). Such immersions in nature have also been proven to bring a host of psychological benefits, as people engage in fewer negative emotions and less self-referential thinking (Bratman et al. 2015).

I consider this a distinct category in my research and teaching, it being a way to be more connected and whole as a human being. This connectedness and wholeness can take either the form of observing, in a focused way, flora or fauna in nature, or a more immersive experience, where you go into nature and just allow yourself to be present to it.

The hypothesis that my colleagues and I have been studying now for six to seven years is that people who experience a greater sense of connection to nature are more likely to care for others and future generations.

We have already spoken about well-being and happiness, and there is a lot of research in that domain. It also changes people's behavior, as I will explain shortly.

There are many studies that show the benefits of nature connectedness, you see four of them in **Figure 4.** The first two are academic studies that covered connectedness to nature and its consequences. The second study, by Capaldi, Dopko, and Zelenski (2014), is actually a meta-analysis. It examined

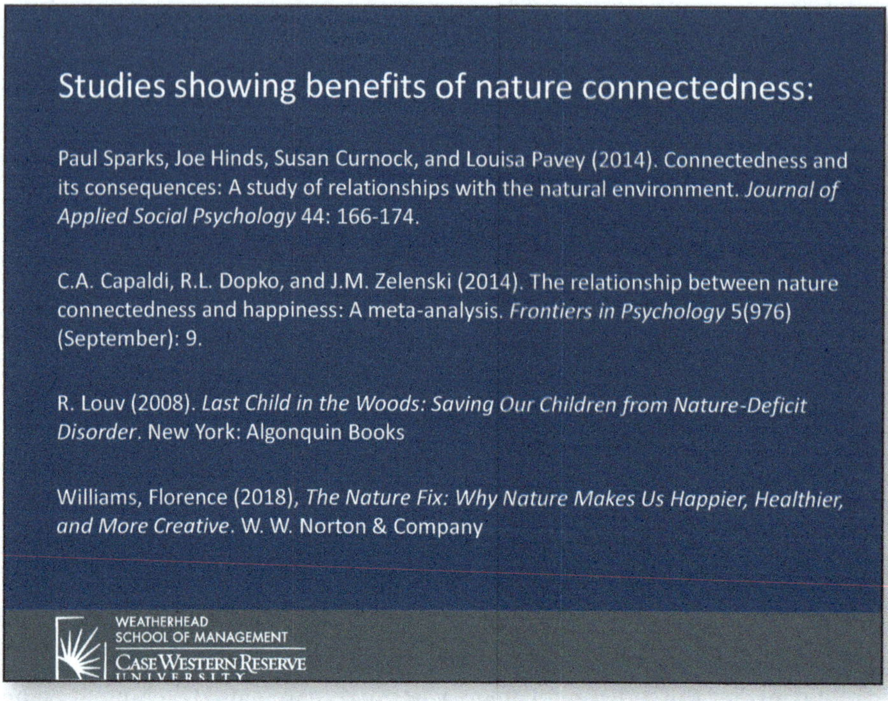

Figure 4

the correlation between nature connectedness and happiness across a great number of different studies and found a statistically significant correlation. A relationship or correlation does not imply causation in this case, but it does show that, generally speaking, more time in nature is associated with a greater sense of happiness and well-being.

The last two are books, with the first being Richard Louv's *The Last Child in the Woods: Saving Our Children from Nature-Deficit Disorder.* The second is a more recent book by Florence Williams, titled, *The Nature Fix: Why Nature Makes Us Happier, Healthier, and More Creative.*

I will now delve into what science says about our experience of nature and how scientists conceptualize our understanding of the experience we have when we connect to nature. To do this, I will review some of the fundamental ideas that we have had in science over the last 300 years, as well as some of the new, emerging ideas in science. Let us start with what others have said about science. Science exercises a huge but often hidden influence in our thinking and acting. This point was made by Thomas Kuhn in his book on scientific revolutions (1962, 2012). Kenneth Gergen (1978) and others have also emphasized this point.

We now turn to some more fundamental ideas in science, so that we can have this interplay between theory and practice—the practice of going into nature. This refers to the practice of immersing ourselves in nature, whether it be through meditation in nature, presence in nature, or the like. If we have that experience once, we can also have a theory that helps us understand what occurred during that practice. That gives us a more complete sense of our connection—human connection—to nature.

We start with physics because it is in some ways the most fundamental of the sciences in describing the behavior of the world we live in. Physics, for much of the last 300 years, has been based on the concepts of particles and forces. Actually, the idea of atoms seen as separate particles and empty space goes all the way back to Greek philosophers like Democritus, who first came up with the term *atomos* in Greek. In **Figure 5**, you see an illustration of two atoms, with their protons and neutrons in the nucleus, and electrons spinning around them at great distances from them, and the atoms are somehow separated in empty space. In such models, the only forces acting on them are gravity, electromagnetic fields, and the strong and weak nuclear forces.

That was the paradigm of the science we grew up with. However, quantum physics in particular is giving rise to a new idea that, in fact, at the finest level of the universe, we have a connected and coherent unified field, a field of energy and information that connects us. Rather than thinking about space as being a vacuum, space, in fact, is a plenum. We now know that the universe contains dark matter, dark energy, and gravitational waves. Along with these fields of energy, vibrational fields of energy connect us not just metaphorically—we are actually connected.

Quantum physicists speak about objects or living systems as excitations of the field quantum physics, which is now a proven domain. Experiments

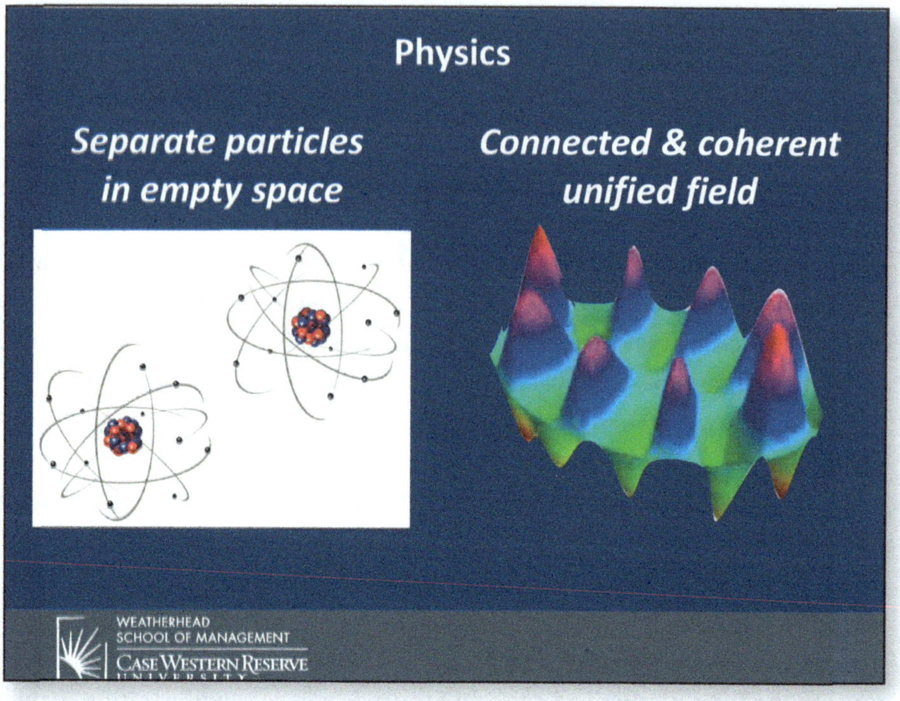

Figure 5 **Left image:** https://m-cacm.acm.org/news/163184-dont-call-it-vaporware-scientists-use-cloud-of-atoms-as-optical-memory-device/fulltext?mobile=true

such as the double-slit experiment, as well as Bell's theorem experiment (**Figure 6**), which I touch on briefly, show this interdependence and describe the fundamental reality that we have.

Bell's theorem experiment was a way to examine the behavior of paired particles. **Figure 6** shows a Source that generates particles V_1 and V_2, which are paired in the sense of having the same wave phase. Once they are paired, they are shot off in opposite directions to crystal A and crystal B, each of which has a mirror in it. Each mirror has a 50% chance of shooting the particle up and 50% chance of shooting the particle down.

This experiment has shown that paired particles that are shot out remain paired even across great distances. Thus, if V_1 and V_2 are paired, and if V_1 hits crystal A and goes up (+1), then V_2 (shot out at the same time) will

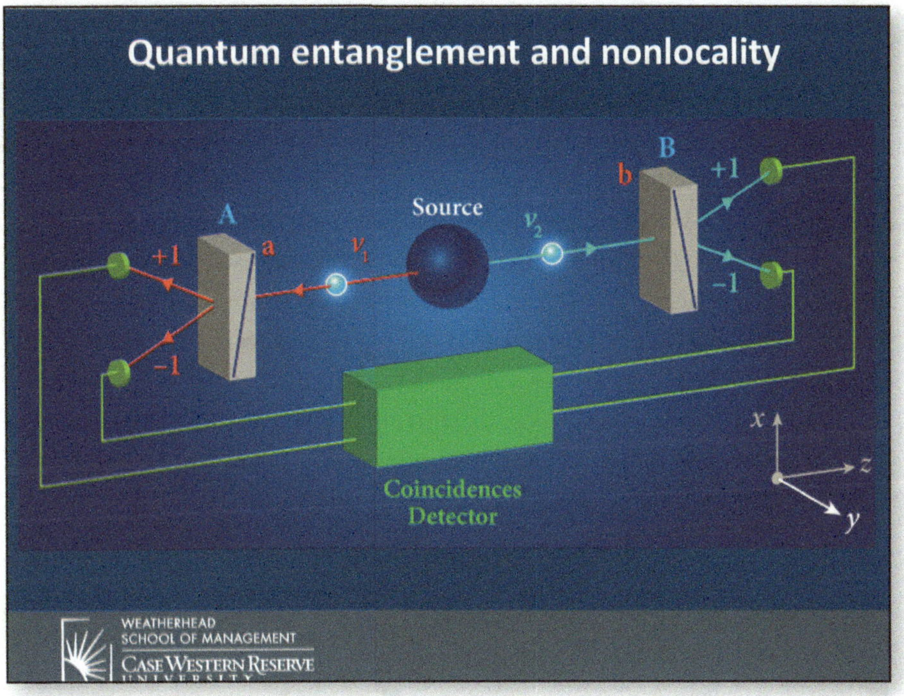

Credit: APS/Alan Stonebraker; Fair Use image **Figure 6**
Source: https://physics.aps.org/articles/v8/123; (https://alanstonebraker.com/)

hit crystal B and also go up (+1). This can be repeated tens of thousands of times, as this is what the coincidences detector shows, and you will never once find the case in which paired particles emerging from the crystals go in opposite directions. This includes cases in which V_1 goes up and V_2 goes down (-1) or the reverse, with V_2 going up and V_1 going down. This is what quantum physicists call entanglement, not nonlocality, and it can happen across very great distances.

That degree of instant correlation, holding over very great distances, suggests that it happens faster than the speed of light. Thus, Erwin Schrödinger, one of the early great quantum physicists, concluded that quantum physics reveals a basic oneness of the universe. I think this is important because, as we start thinking about our relationship with nature, we start seeing that at this most fundamental level, we are one with the world around us.

We also have other sciences, such as epigenetics, that tell us that it is not only the genes we are born with that determine things such as life expectancy and diseases we may have in life, but also our relationship with nature. What that relationship is, whether we are exposed to pollution or to healthy nature, affects the proteins that wrap the genes and lead to gene expression, either good or bad.

Finally, from the nineteenth century onward, the rise of Darwinism and then neo-Darwinism, as well as the economics of William Stanley Jevons and John Stuart Mill, led us to believe that human beings were essentially selfish, competitive, and separate—what the existentialist called the "bounded human being." We are more recently starting to see that there are, in fact, lessons from nature by which we can better understand human nature as relational, cooperative, and connected.

Consciousness also is undergoing an interesting controversy, with phys-

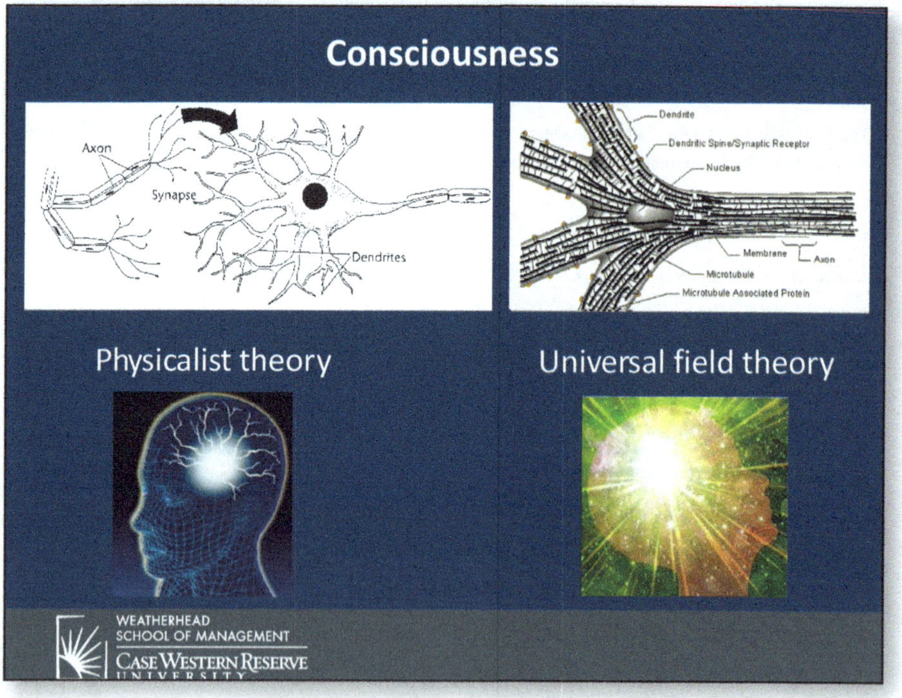

Figure 7

Left image top: http://science.halleyhosting.com/sci/soph/anatomy/nervous/neuron.htm
Left image bottom: https://ruayres.wordpress.com/2014/10/12/working-with-exernomics/
Right image top: https://www.tony5m17h.net/QuantumMind2003.html
Right image bottom: https://www.tsemrinpoche.com/tsem-tulku-rinpoche/etc/the-benefits-of-pineal-gland-activation.html?nomobile

icalist theory versus universal field theory. Physicalist theory maintains that we generate consciousness just inside our brain, like a supercomputer, while universal field theory suggests that consciousness is actually a property of the universe that we can tap into.

Thus, our brains are almost instruments that tap or tune into a universal consciousness through microtubular lattices (**Figure** 7), and there is good scientific research emerging on this.

All of the science now tends to converge with spiritual traditions, illustrated in **Figure 8**. For example, in the teachings of the Vedanta from the Hindu tradition, or Vedic tradition, starting with the Rig Vedas and then the Upanishads, there has always been this idea of "Brahma," the background field from which the manifest universe comes. On the right side of the figure, you see Native American traditions that consider human beings as relatives of animals and plants. "All My Relatives" refers to how a Native American

Figure 8

Ubuntu: https://www.facebook.com/WilburSmith/photos/ubuntu-is-a-south-african-term-meaning-humanity-it-is-often-translated-as-i-am-b/4030183166996687/?_rdr
All my relatives: https://www.amazon.com/All-My-Relatives-Exploring-Indigenous/dp/080329994X
Dao: https://www.sammyboy.com/threads/want-to-be-successful-avoid-toxic-co-workers-and-culture.271847/page-3
Buddha: public domain

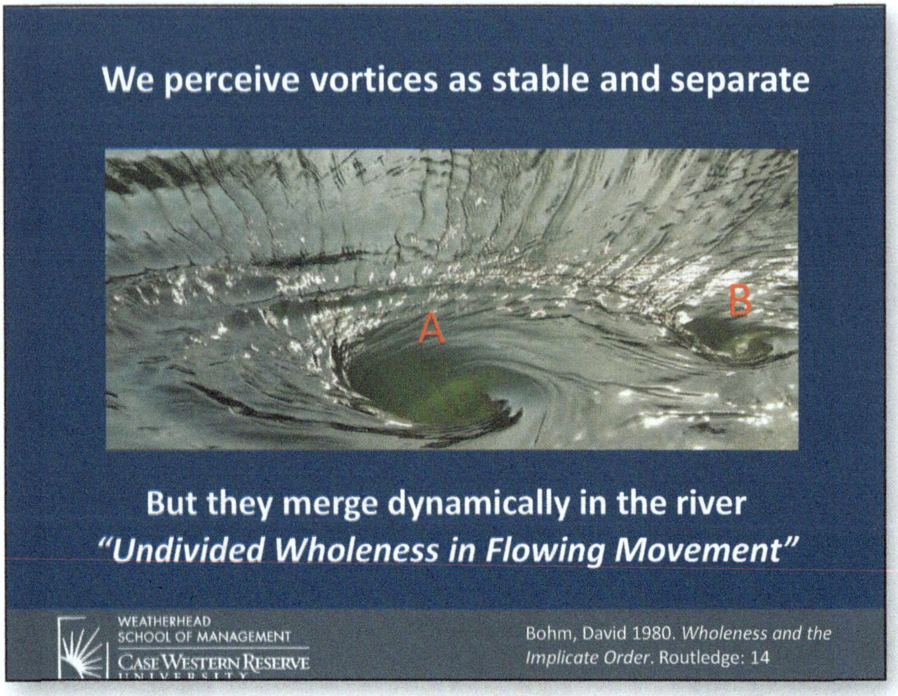

Figure 9 **Source:** https://www.shutterstock.com/video/search/green-whirlpool

would see a rabbit or fox or deer or even a tree.

In Africa, you have Ubuntu, the idea that "I am who I am because of who we all are." In China, Japan, and elsewhere, we have Taoism (Daoism), the notion that there is the "Way" (Dao) and that in practice we can become one with the Way. In Buddhism as well, he who experiences the unity of life sees his own Self in all beings and all beings in his own Self. In the center, you have Ein Sof, which is an ancient symbol from the Kabbalah, an early thread in Judaism, that also has this notion of an ineffable background to the reality you experience.

In conclusion, we tend to see the world as separate objects, like separate vortices in a river. **Figure 9** shows a river in which vortex A and vortex B seem stable and separate, as if they have their own structure in time and space. Perhaps a better way to see them is how they are merging dynamically in the river. David Bohm, the quantum physicist, called it "undivided wholeness in flowing movement."

I would like to finish by noting some benefits of connecting to nature (Beck 2017), including in my own research and in a lot of the research that has already been done. Some of the benefits that have been highlighted in contemporary research on this practice include: overall health improvement (Gesler 1992); stress relief (Wells and Evans 2003); reduced negative emotions, such as decreased fear and anger; enhanced positive effects; improvements in mood and increased subjective well-being (Mayer and Frants 2004); feelings of joy and happiness (Brereton, Clinch, and Ferreira 2008); a sense of reconnection with self (Korpela 1989); kinship ties in teams (Beverly and Whittemore 1993); a heightened sense of community, kinship, egalitarianism, and belongingness, along with increased empathy (Florence 2017); a stronger sense of place (Casey 1998); and improved cognitive abilities, including creativity, cognitive flow, and mental performance in problem solving (Chalquist 2009).

Connecting to nature also increases personal well-being, it can raise awareness of how our actions impact others, and it can transform people and leaders, in particular, by increasing their emotional, social, and spiritual intelligence. It can increase entrepreneurial creativity and collaboration, and perhaps, very importantly at this time in human history, can strengthen pro-social and pro-environmental behaviors.

In the business courses I teach on flourishing enterprise, it is becoming clear that flourishing in business requires both a strong financial business case and behavioral change, and that the consciousness of connectedness, including a consciousness and connectedness to nature, are central to lasting behavioral change.

With this, I conclude my presentation, and I look forward to the panel. Thank you. ◊

Acknowledgments

This paper's content draws on selected works of Albert Einstein, Niels Bohr,

John Archibald Wheeler, Erwin Schrödinger, Wolfgang Pauli, David Bohm, Richard Feynman, Werner Heisenberg, and Max Planck. It benefits greatly from pioneers who sought to integrate quantum physics and philosophy: David Bohm's *Wholeness and the Implicate Order* (1980), Ken Wilber's *Quantum Questions* (1984); Margaret Wheatley's *Leadership and the New Sciences* (1994); Ervin Laszlo's *The Interconnected Universe* (1995) and *The Self-Actualizing Cosmos* (2014); among others. I owe a further debt to Paul Levy's *Quantum Revelation: A Radical Synthesis of Science and Spirituality* (2018), Lothar Schäfer's *Infinite Potential: What Quantum Physics Reveals About How We Should Live* (2013), and Allan Combs' *Consciousness Explained Better: Towards an Integral Understanding of the Multifaceted Nature of Consciousness* (2009).

Revisiting these works was part of a broader research program on Quantum Leadership at Case Western Reserve University, starting in 2014 and funded by the businessman and philanthropist Fred Chavalit Tsao. Extensive field research led to our book, *Quantum Leadership: New Consciousness in Business,* published by Stanford University Press in 2019.

References

Beck, J. 2017. "Nature Therapy Is a Privilege." *ICEWIM*. Accessed on July 12, 2022. https://www.icewim.co/post/nature-therapy-is-a-privilege

Beverly, E.A., and R.D. Whittemore. 1993. "Mandinka Children and the Geography of Well-Being." *Ethos* 21 (3): 235–272. http://www.jstor.org/stable/640550

Bohm, D. 1980. *Wholeness and the Implicate Order.* England, UK: Routledge.

Bratman, G.N., J.P. Hamilton, K.S. Hahn, G.C. Daily, and J.J. Gross. 2015. "Nature experience reduces rumination and subgenual prefrontal cortex activation." *Psychological and Cognitive Sciences* 112 (28): 8567–8572. https://doi.org/10.1073/pnas.1510459112

Brereton, F., J.P. Clinch, and S. Ferreria. 2008. "Happiness geography and the environment." *Ecological Economics* 65 (2): 386–396. https://doi.org/10.1016/j.ecolecon.2007.07.008

Capaldi, C.A., R.L. Dopko, and J.M. Zelenski. 2014. "The relationship between nature connectedness and happiness: a meta-analysis." *Frontiers in Psychology* 5: 976. https://doi.org/10.3389/fpsyg.2014.00976

Carson, R. 1962. *Silent Spring.* Boston, MA: Houghton Mifflin Company. Cited in Suzuki (1999). *The Sacred Balance: Rediscovering our Place in Nature,* 221. Vancouver, BC: Greystone Books.

Casey, E.S. 1998. *The Fate of Place: A Philosophical History.* Berkeley, CA: University of California Press. Cited in Chalquist (2009).

Chalquist, C. 2009. "A Look at the Ecotherapy Research Evidence." *Ecopsychology* 1 (2): 64–74. https://doi.org/10.1089/eco.2009.0003

Combs, A. 2009. *Consciousness Explained Better: Towards an Integral Understanding of the Multifaceted Nature of Consciousness.* St. Paul, MN: Paragon House.

Gergen, Kenneth. 1978. "Toward Generative Theory." *Journal Of Personality And Social Psychology* 36 (11): 1344–1360. DOI: 10.1037//0022-3514.36.11.1344. https://works.swarthmore.edu/fac-psychology/326

Gesler, W. 1992. "Therapeutic landscapes: Medical issues in light of the new cultural geography." *Social Science Medicine* 34 (7): 735–746. https://doi.org/10.1016/0277-9536(92)90360-3

Korpela, K.M. 1989. "Place-Identity as a product of environmental self-regulation." *Journal of Environmental Psychology* 9: 241–256. https://doi.org/10.1016/S0272-4944(89)80038-6. Cited in Chalquist (2009).

Kuhn, T. 1962. *The Structure of Scientific Revolutions.* 50th Anniversary Edition (2012). Chicago: The University of Chicago Press.

Laszlo, E. 1995. *The Interconnected Universe: Conceptual Foundations of Transdisciplinary Unified Theory.* Singapore: World Scientific Publishing Company.

———. 2014. *The Self-Actualizing Cosmos: The Akasha Revolution in Science and Human Consciousness.* Rochester, VT: Inner Traditions.

Levy, P. 2018. *Quantum Revelation: A Radical Synthesis of Science and Spirituality.* New York, NY: SelectBooks, Inc.

Louv, R. 2008. *Last Child in the Woods: Saving Our Children from Nature-Deficit Disorder.* New York City, NY: Algonquin Books.

Mayer, S.F., and C. McPherson Frantz. 2004. "The connectedness to nature scale: A measure of individuals' feeling in community with nature." *Journal of Environmental Psychology* 24 (4): 503–515. https://doi.org/10.1016/j.jenvp.2004.10.001

Park, B.J., Y. Tsunetsugu, T. Kasetani, T. Kagawa, and Y. Miyazaki. 2010. "The physiological effects of Shinrin-yoku (taking in the forest atmosphere or forest bathing): evidence from field experiments in 24 forests across Japan." *Environmental Health and Preventive Medicine* 15: 18. https://doi.org/10.1007/s12199-009-0086-9

Schäfer, L. 2013. *Infinite Potential: What Quantum Physics Reveals About How We Should Live.* Deepak Chopra.

Sparks, P.J. Hinds, S. Curnock, and L. Pavey. 2014. "Connectedness and its consequences: a study of relationships with the natural environment." *Journal of Applied Social Psychology* 44 (3): 166–174. https://doi.org/10.1111/jasp.12206

Tsao, F.C., and C. Laszlo. 2019. *Quantum Leadership: New Consciousness in Business.* Stanford, CA: Stanford University Press.

The National Trust. 2016. "The humble stick revealed as the must-have toy for summer." *The National Trust.*

Wells, N., and G. Evans. 2003. "Nearby nature: A buffer of life stress among rural children." *Environment and Behavior* 35 (3): 311–330. Cited in Chalquist (2009).

Wheatley, M.J. 1994. *Leadership and the New Science: Discovering Order in a Chaotic World.* Third Edition. San Francisco, CA: Berrett-Koehler Publishers, Inc.

Wilber, K. 1984. *Quantum Questions: Mystical Writings of the World's Great Physicists.* Boston, MA: Shambhala Publications, Inc.

Wild Ozark. 2017. "What is Nature Immersion? It's NOT being immersed in natural disasters." *Wild Ozark.* Accessed on July 12, 2022. https://www.wildozark.com/nature-immersion/

Williams, F. 2017. *The Nature Fix: Why Nature Makes Us Happier, Healthier, and More Creative.* New York, NY: W.W. Norton & Company.

Wilson, E. O. 2009. *Biophilia.* Reprint Edition. Cambridge, MA: Harvard University Press. 1–2.

Session 3 | First Commentary
on the Presentation by Dr. Chris Laszlo

Exploring the Connections between Humans and Nature

Dr. Julia Storberg-Walker
Chair, Department of Human and Organizational Learning,
The George Washington University, Washington, D.C., USA

I am really pleased to be invited to talk about something I am really excited about.

I love everything about the image in **Figure 1**. I love the trees, the branches, the leaves, the moving water, and the still water in the eddies. I like the glistening water on the rocks, and I can almost feel the moss on them. There is something special about our human connection to nature.

This is what Dr. Laszlo spoke about in his presentation (**Figure 2**). It is not an abstract thing—it is a felt sense. It is a feeling of your breath slowing down and a feeling of peace, perhaps descending into your bones. It is also so curious—what is this experience? Dr. Laszlo, my friend Chris, talked

About the Speaker

- Chair and Associate Professor,
  Department of Human and Organizational Learning;
  Director, Organizational Leadership and Learning Master's Degree Program,
  The George Washington University, Washington, D.C., USA.
- Former Editor-in-Chief, *Human Resource Development Review.*
- Transdisciplinary activist/scholar work grounded in new materialism, posthumanism, quantum field theory, and wisdom traditions spanning diverse cultures and historical moments.
- PhD in Work, Community, and Family Education, University of Minnesota

Figure 1 **Source:** https://www.pexels.com/photo/waterfalls-in-forest-460621/

about a quantum understanding of humans in nature.

He asked the really important question, "How does science conceptualize our understanding of the experience when we connect to nature?" I think this is a very profound question. It is a question that probes deep into what it means to be human, and it asks us to look at science with curiosity and not with certainty. The question also challenges us to consider that there may be different answers to scientific questions. In this question, it is about how and why humans experience nature.

Like Dr. Laszlo, I have been deeply immersed in learning about quantum theory, and the implications of quantum ontology for education, learning, and teaching. The more I found out about quantum theory and quantum ontology, the more I saw connections to the spiritual and religious texts that I had been studying for years, as well as connections between the scientific

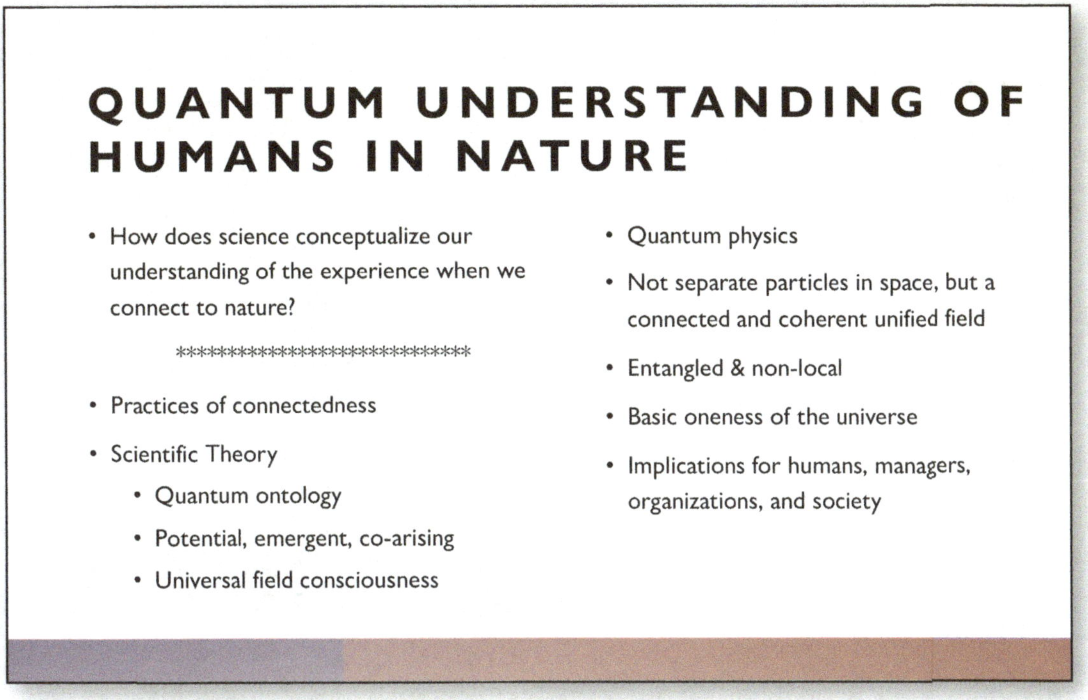

Figure 2

ideas in the quantum paradigm to those perennial wisdom ideas found in those diverse religious texts. Most importantly, it was my personal experiences with what Dr. Laszlo calls "direct knowing" that compelled me to lean into quantum ideas, and to start voraciously reading all I could find out about connections between quantum theory, spirituality, religion, and science.

As you know, Dr. Laszlo conducted pathbreaking research on leaders who have had profound shifts in consciousness. His study found that the key to making this shift is to somehow have an experience of oneness, a moment of clarity that the entire cosmos is unified, with the weird new awareness that being human is being infused within nature, not separate from it. Today, he described the scientific evidence that shows the positive benefits of spending time in nature. We see here, biologically and psychologically, study after study shows how human bodies and moods are improved by spending time in

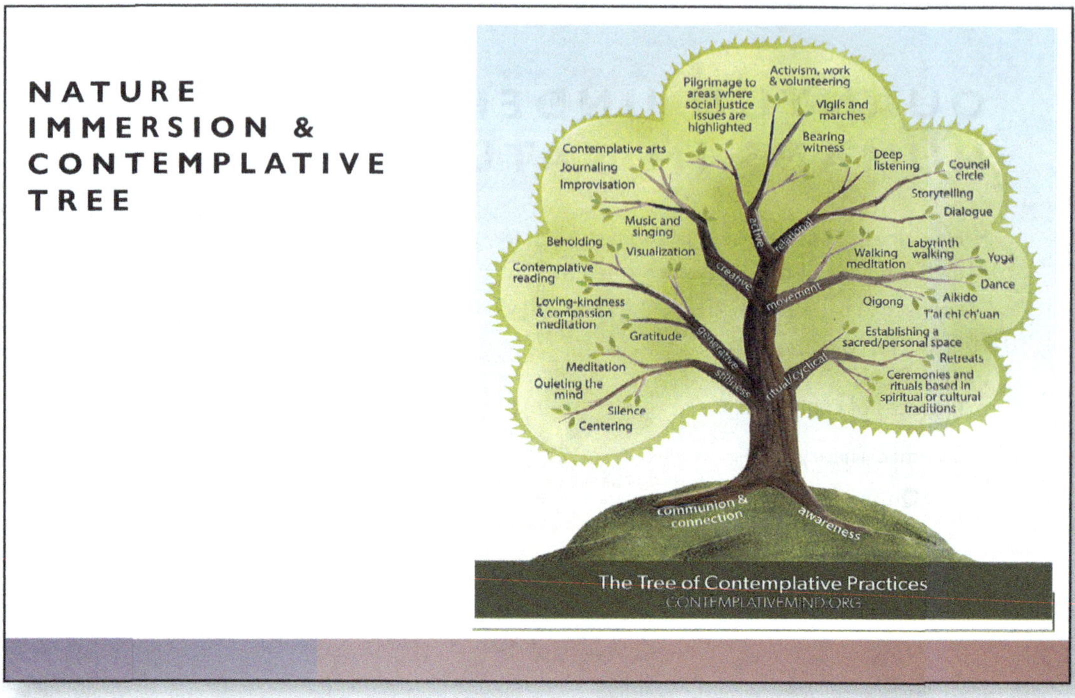

Figure 3 **Image source:** https://www.contemplativemind.org/practices/tree

nature and other contemplative practices, such as those shown in **Figure 3**.

Why is this so? As Dr. Laszlo described it, one scientific way to think about this would be to look at the neuronal and chemical answers to the question. This is a traditional and materialist view. It is more traditional, and the research and results would focus on the body parts that send and receive information relating to mood and physical well-being.

Science has contributed greatly to humankind. It can contribute to our well-being and broaden what is possible. Science also has allowed us to see ourselves in a new way. The earthrise image in **Figure 4** was made possible only through the steady advances in science over the decades. Interestingly, it also started a movement that started to decentralize normal, traditional science as the only arbiter of knowledge and propel the emergence of the environmental movement, and that made the way for traditional, indigenous

Image source: https://pixabay.com/photos/earth-earth-rise-moon-moon-surface-11038/ **Figure 4**

science and wisdom to start to reclaim its place in our knowledge structures.

The quantum scientific perspective was the focus of Dr. Laszlo's presentation. To understand human relations with nature, a person informed by the quantum perspective would look to the larger field of energy or information for answers. Building off of both theoretical and empirical quantum research, he made the claim that humans are not separate from the research objects that they study. He also said that human consciousness is "a fundamental property of all life, and irreducible to physical phenomena."

To me, the most critical ideas Dr. Laszlo raised were focused on quantum theory elements, such as entanglement, co-emergence, and nonlocality, which all have implications for human experiences of nature on this planet. Because we humans are in nature, and because nature from the quantum view seeks wholeness or unity, there is a reciprocal relationship between

Figure 5 **Image source:** https://www.pexels.com/photo/person-holding-container-with-seaweed-2280568/

humans and nature. Cooperation and collaboration, rather than competition, are guiding principles of how we live and who we are in relationship with. All of these influence our health and well-being, such as through epigenetics. Ultimately, all living systems, he argues, are dynamically connected to each other and to the larger ecosystem.

Why is this important? **Figure 5** illuminates what I think are two sides of the same coin of science. One side contributes to human well-being, and the other side diminishes human well-being. In **Figure 5**, the scientist is separate from nature, as you can see the scientist holding nature in a petri dish, looking at it from the outside, and performing an action onto it.

The scientist is conducting experiments on nature, rather than experiments with nature. To me, this illuminates that science has a partial view. This partial view, the traditional or materialist view, does not see the deeper

Left image source: https://www.theguardian.com/environment/2016/nov/17/whats-air-pollution-like-where-you-are-share-your-experiences **Figure 6**
Photograph: Harish Tyagi/EPA, or https://www.bbc.com/future/article/20200813-the-hidden-harm-of-medical-plastic-waste-and-pollution (image: EPA) **Right image source:** https://www.pexels.com/photo/drone-shot-of-dumpsite-3174350/

connections humans have with nature. Nor does the traditional, scientific view accept other ways of knowing, such as indigenous science or intuition. In a sense, I see that we humans are metaphorically in the petri dish with this object—it completely flips the paradigm.

We are a part of nature. The downside that we have not grappled with is how we do not understand we are facing a crisis because we do not think we are a part of nature.

Let me describe just one industry, with health care as an example. If health care was a country, it would be the fifth largest emitter of greenhouse gases in the world (Karliner et al. 2019). One of the big problems of health care is that the frontline worker with the patient really wants to just do what is right for that patient, without considering reuse or recycling. This focused view on the patient disregards the downstream effects of all the plastics and materials used. This view of focusing on the small and not really looking at the big is detrimental to the larger ecosystem, as you can see in **Figure 6.**

What Dr. Laszlo argues, and what I support, is that we need to build new capacities in ourselves, new skills and new ways of being. People like him are working tirelessly to catalyze this shift in consciousness.

This is supported by theoretical and empirical quantum research, and it is made possible through practices such as nature immersion. It is also a remembering of indigenous ways of knowing and being. Science is shifting and coming closer to ancient wisdom traditions, from multiple cultures. God, nature, and the cosmos are being made accessible through quantum ideas.

Thank you for your attention and for your intention to help catalyze this shift in consciousness in all of us. ◊

References

Karliner, J., S. Slotterback, R. Boyd, B. Ashby, and K. Steele. 2019. "Health Care's Climate Footprint: How the Health Sector Contributes to the Global Climate Crisis and Opportunities for Action." *Health Care Without Harm, in collaboration with ARUP.* 19. Accessed on July 13, 2022. https://noharm-uscanada.org/ClimateFootprintReport

Session 3 | Second Commentary
on the Presentation by Professor Chris Laszlo

Quantum Physics and Associated Connections between Humans and Nature

Dr. Glen Rein
Quantum-Biology Research Lab, Boston, USA

The theme for this session is, "Frontier Science Perspectives on Our Relationship with the Natural World." To understand our relationship with the natural world, we should consider the energetic side of nature and consider our consciousness as our energetic essence.

Unfortunately, many neuroscientists think of consciousness as a byproduct of our physical brain, which is considered fundamental. Nonetheless, all will agree that the human body is electrochemical in nature and that electrical energy is a critical part of our constitution. However, this intrinsic

About the Speaker

- Founder, Quantum-Biology Research Lab, Boston, USA. Funded by the Fetzer Institute, the lab has pioneered the study of biological effects of nonclassical electromagnetic (EM) fields
- Discovered and published a new effect of neuropeptides on immune cell function
- Performed traditional biomedical research at Harvard and Stanford Medical Schools, involving studies in neuroscience, immunology, psychoneuroimmunology, and bioelectromagnetics
- Published in *Nature* the first demonstration that EM fields modulate neurotransmitter function *in vitro*
- Co-developed a new electro-diagnostic method to measure endogenous EM fields of the body
- PhD in Neurochemistry; PhD in biochemistry from the University of London

bioenergy is not limited to electrically excitable cells in the brain and heart; it is present in every cell in our bodies. Thus, considering consciousness as a fundamental type of bioenergy, we can conclude that consciousness is present in all our cells.

In his presentation on quantum physics and associated connections between humans and nature, Dr. Laszlo provided scientific evidence for the benefits of being connected with nature. They include stress relief, reduction in negative emotions such as fear or depression, and enhanced positive emotions such as happiness, kinship, egalitarianism, and empathy. Connecting to nature also has a beneficial effect on human behavior since we experience more kinship with others and a heightened sense of community. Such humanitarian behavior is clearly needed in today's hectic world with pandemics and war.

In addition to psychological and emotional benefits of being connected with nature, Dr. Laszlo also discussed the effect nature has on the human brain and mind. Nature also has a beneficial effect on our cognitive abilities, including creativity and problem solving. Furthermore, immersion in nature results in lowering sympathetic nerve activity and increasing parasympathetic activity of the peripheral nervous system. These neuronal effects were observed by Japanese researchers studying the relaxing effects of "forest bathing," which is ultimately controlled by the brain, and specifically by lowering the stress hormone cortisol.

Nature also induces the experience of oneness and brings out our compassion, or "goodness." From a neurophysiology perspective, we may consider the role of the superior parietal lobe in the frontal cortex. This brain region does much more than helping us visualize our internal and external worlds. The parietal lobe also impacts how we relate to space and the environment, as well as humans in our environment.

Recently, from a neurotheology perspective, Arion (2017) pointed out that the parietal lobe can help us feel more connected to something beyond ourselves by appreciating our connection to the oneness of the Universe. This expanded sense of self often occurs in mindfulness meditation practices, which is also now known to involve the parietal lobe (Marchand 2014). Such meditation practices allow the individual to access their consciousness.

Dr. Laszlo further pointed out that quantum physics is providing the new concept that, "at the finest level of the universe," there is a "connected and coherent" unified field of energy and information that connects us.

Parallels between the quantum aspects of human consciousness and quantum aspects of the Universe are abundant in the scientific literature. Thus, resonance between the quantum fields in the human body and quantum fields in nature offer a mechanism for how nature can affect humans and how humans can affect nature. However, the fundamental question still arises as to how quantum fields in the brain, whether intrinsic or exogenous (that is, in nature), can affect our emotions, our cognitive functions, and even our behavior as so elegantly pointed out by Dr. Laszlo.

One possible explanation for how quantum fields affect neurochemical processes in the brain has been proposed (Rein 2016). This theory suggests that subtle quantum fields are down-converted and made more dense via various transformations in the brain (Rein 2017). One such transformation is a process called reverse phase conjugation. Phase conjugation is a process whereby ordinary light (albeit coherent, laser light) can be up-converted to a quantum field after reflecting off a phase conjugation mirror (Bloom and Bjorklund 1997). The reflected energy is described as "time-reversed" and "virtual" in the physics literature.

The reverse of this process has also been described, where a quantum field can be down-converted into coherent electromagnetic light. Electromagnetic fields are well-known to produce a variety of neurochemical changes in the brain (Gunderson, Greenebaum, and Schaller 1986), including direct effects on neurotransmitters (Rein and Dixey 1984).

Although phase conjugation has been proposed as a mechanism involved in memory (Marcer and Schempp 1997), it was never clear where the phase conjugation mirror was located in the brain, or if any biological system exhibited the phase conjugation phenomenon. However, Devane (1984) discovered that porphyrin molecules, which are associated with glycoproteins in the plasma membrane of all cells, exhibit phase conjugation behavior. Thus, it is conceivable that phase conjugation can occur in the brain and that quantum fields in nature could enter the brain and be down-converted into electromagnetic fields and thereby influence human emotions and behavior. ◊

References

Arion, A.-C. 2017. "A NEW CONCEPT TO BE RECKONED WITH: NEUROTHEOLOGY." In I*nternational Multidisciplinary Scientific Conference on the Dialogue between Sciences & Arts, Religion & Education* 1: 22–32. Ideas Forum International Academic and Scientific Association. Accessed on July 13, 2022. https://web.archive.org/web/20220227185157/https://ifiasa.org/gallery/2017-2.2nd-2017mcdsare-arionalexadru.pdf

Bloom, D.M., and G.C. Bjorklund. 1977. "Conjugate wave front generation and image reconstruction by four-wave mixing." *Applied Physics Letters* 31: 592–598. https://doi.org/10.1063/1.89791

Devane, M.M. 1984. "Temporal investigation of phase conjugation, with enhancement, in magnesium tetraphenyl porphyrin (MgTPP)." *Optics Communications* 52 (2): 136–140. https://doi.org/10.1016/0030-4018(84)90298-0

Gundersen, R., B. Greenebaum, and M. Schaller. 1986. "Intracellular recording during magnetic field application to monitor neurotransmitter release events: Methods and preliminary results." *Bioelectromagnetics* 7 (3): 271–281. https://doi.org/10.1002/bem.2250070304

Marcer, P.J., and W. Schempp. 1997. "Model of the Neuron Working by Quantum Holography." *Informatica* 21: 519–534.

Marchand, W.R. 2014. "Neural mechanisms of mindfulness and meditation: evidence from neuroimaging studies." *World Journal of Radiology* 6 (7): 471–479. http://dx.doi.org/10.4329/wjr.v6.i7.471

Rein G. 2016. "A quantum-chemical approach to consciousness based on phase conjugation." *Cosmos and History: The Journal Natural and Social Philosophy* 12: 250–258. Accessed on July 13, 2022. https://www.thefreelibrary.com/A+quantum+chemical+approach+to+consciousness+based+on+phase...-a0471144311

———. 2017. "Mind as a virtual phase-conjugated hologram." *Cosmos and History: The Journal Natural and Social Philosophy* 13 (2): 217–226.

———, and R. Dixey. 1984. "Neurotransmitter Release Stimulated in a Clonal Nerve Cell Line by Low Intensity Pulse Magnetic Fields." In *Nonlinear Electrodynamics in Biological Systems,* edited by W.R. Adey and A.F. Lawrence. Boston, MA: Springer. https://doi.org/10.1007/978-1-4613-2789-9_6

Session 3 | Presentation 2

A Vision for Earth's Future Arising from Frontier Science

Professor Gary E. Schwartz
Psychology, Medicine, Neurology, Psychiatry, and Surgery,
The University of Arizona, Tucson, USA

All matter originates and exists only by virtue of a force... We must assume behind this force the existence of a conscious and intelligent Mind. This Mind is the matrix of all matter.

—Max Planck

Clearly, the universe had meaning and direction ...an unseen dimension behind the visible creation that gives it an intelligent design and that gives life purpose.

—Edgar Mitchell

About the Speaker

- Professor of Psychology, Medicine, Neurology, Psychiatry and Surgery, The University of Arizona, Tucson
- Director, Laboratory for Advances in Consciousness and Health
- Founding President, Academy for the Advancement of Postmaterialist Sciences
- Areas of expertise include: alternative medicine and healing; consciousness and life after death; integrative medicine; psychophysiology; spiritual psychology
- Books authored and coauthored include: *The G.O.D. Experiments* (2007); *The Sacred Promise* (2011); *Super Synchronicity* (2017); *Greater Reality Living* (2018); *Extraordinary Claims Require Extraordinary Evidence* (2021)
- PhD in Psychology, Harvard University

The title of this invited presentation, *A Vision for Earth's Future Arising from Frontier Science,* was requested by the organizers of the Third International Conference on Science and God (ICSG III), with the theme Environmental Restoration in the Era of Frontier Science. Here is how the conference organizing committee described the intended scope of this presentation:

> *If humans can live in harmony with nature, treating nature as sacred and expressing divine love toward all created things, then the Earth's future will be very bright. This talk will show us how frontier science can help us in this process.*

A few of my books speak to the scope proposed by the organizing committee. They include:

- *The G.O.D. Experiments: How Science is Discovering God in Everything, Including Us* (Schwartz 2006), which includes an Epilogue titled, "Infinite Love: The Ultimate Gift from God?" (G.O.D. stands for Guiding, Organizing, Designing process / intelligence.)
- *The Sacred Promise: How Science is Discovering Spirit's Collaboration with Us in Our Daily Lives* (Schwartz 2011), which includes an Afterword titled, "Spirit is Calling Us."

It should not come as a surprise that I paid close attention to the organizing committee's words "sacred" and "divine love toward all created things." To explicitly honor the chosen title of these international conferences—namely, Science and God—and in light of the time and space limitations for this presentation, I will focus my comments on one emerging area of frontier science that speaks directly to the "brightness of the Earth's future," the controversial perspective of emerging divine science, what Matzke and Tiller (2020) call "Source Science" in their book *Deep Reality: Why Source Science May Be the Key to Understanding Human Potential.*

How can Source Science, and its sister science, what I think of as Soul

Science, "help us with this process?" Let us briefly consider some of the frontier evidence for Source Science and Soul Science before considering its seismic implications for society.

Is the "One Mind" a Universal "Guiding, Organizing, Designing" Mind?

Many authors have written about the emerging sciences of source and soul. Some scientists come from parapsychology, as illustrated by the visionary book written by Luis Portela, chairman of the Bial Foundation that focuses on funding research in parapsychology and psychophysiology, titled, *The Science of Spirit: Parapsychology, Enlightenment and Evolution* (2021). Others come from more spiritually focused science, as illustrated by the seminal book written by Ervin Laszlo, PhD, a systems scientist and musician, titled *Science and the Akashic Field: An Integral Theory of Everything* (2nd Edition, 2007).

As Larry Dossey makes the case in his book titled, *One Mind: How Our Individual Mind is Part of a Greater Consciousness and Why it Matters* (2014), there is good reason for us to believe that levels of consciousness exist beyond individual minds in people and animals, and the highest levels of mind reach all the way to the hypothesis of a Universal Mind—what Native Americans call "the Great Spirit," what people of Judeo-Christian faith call "God," what scientists like Planck call the "Mind that is the matrix of all matter," and what explorers like Mitchell reference as the "unseen dimension" that gives the Earth and the cosmos "an intelligent design and gives life purpose." The foundational source science question is, where is the evidence that documents the presence of such an active, unseen, super intelligent, and caring, super Mind?

The conclusion I have reached is that some of the best evidence for the existence of such a super Mind comes from the reality of what I call super synchronicities, an extension of Carl Jung's concept of synchronicity. Super synchronicities (or supersynchronicities) are patterns of six or more serial coincidences (or co-occurrences) observable in real life that cannot be explained in simple statistical, psychological, or materialist ways.

I first revealed this evidence and its statistical analysis in my book, *Super Synchronicity: Where Science and Spirit Meet* (2017). I published some compelling exemplars of super synchronicities in a trilogy of papers in the American Psychological Association journal, *Spirituality in Clinical Practice* (Schwartz 2014, 2015a, 2015b). I subsequently included a chapter devoted to super synchronicity in the first volume published by the Academy for the Advancement of Postmaterialist Sciences titled, *Is Consciousness Primary* (Schwartz, Woollacott, and Schwartz 2020). I will briefly outline an exemplary super synchronicity adapted from Schwartz 2015a.

However, before examining this prototypic super synchronicity, we need to keep in mind two deep questions in the spirit of Matzke and Tiller's *Deep Reality* (2020) book:

> **Question 1:** Must we infer the existence of some sort of super intelligent Mind to explain the structure of these stunningly improbable (by random or chance processes) patterns of events in the real world?
>
> **Question 2:** If so, must we infer that the super intelligent Mind can invisibly influence the physical world, including our minds and consciousness, to produce such temporally and spatially organized patterns of events in the real world?

Based on the totality of the emerging evidence, my conclusion regarding both questions is, "Yes." Furthermore, my sense of the emerging evidence is that humanity has the innate capacity to receive guidance and wisdom from this Source under the "right" conditions (for example, when it is sought for the best and highest good of humanity, nature, and the planet as a whole).

Consider the unlikely pairing or co-occurrence of two normally independent things: (1) giraffes, and (2) Paris. It is extremely rare to witness content and images involving both giraffes and Paris in the same context. For example, how often have you seen a large poster of the Eiffel Tower in the same general location as a sculpture of a giraffe that is more than 12 feet high

CO-OCCURRENCE EVENTS	DATE	NATURE OF GIRAFFE - PARIS COINCIDENCE EVENTS
1	6/16	Mr. D. secretly requests Giraffe and Paris information from mediums
2	6/18	G. Schwartz notices Giraffe and Eiffel Tower at Canyon Ranch
3	6/22	G. Schwartz reads Giraffe and French in Davids and Schwartz book
4	6/22	R. Schwartz sees Giraffe picture and reads Paris in My Friend Michael book
5	6/24	Dr. G. sends Giraffe photo and Schwartz's watch two movies involving Paris
6	6/24	Bag displayed with 2 Giraffes, two stores feature Paris name
7	6/24	Novelty store has Giraffe with French writing, Paris store around corner
8, 9, 10, 11	6/26	Four Columbia students share recent Giraffe–Paris coincidences
12	6/26	Drug store features Giraffes and whole window of Paris products
13	6/27	Miller's son wears Giraffe socks, Dr. Parisetti emails me a research proposal
14	6/28	NY Airport, CNN commercial shows Giraffe, then shows Eiffel Tower
15	6/28	Atlanta Airport, store has miniature Giraffe with French writing, Crepe store, woman with luggage that says Paris and purchased in Paris

Table 1 (adapted from Schwartz 2015a). 15 giraffe–Paris co-occurrences experienced within a 12-day period.

(Event 2 in **Table 1**)? Or, how often have you been in an American airport where you spent some time in a store that had miniature giraffes with French writing, and then found yourself standing directly behind a woman with luggage that said Paris, and the woman shared with you that she had bought the luggage in Paris (Event 15 in **Table 1**)?

Most giraffe–Paris co-occurrences are highly improbable. Now, let us imagine that you (plus others you were with) have just experienced a total of 15 giraffe–Paris co-occurrences in a 12-day period (listed in **Table 1**).

The statistical question is, what is the conditional probability that this super synchronicity category of rare co-occurrences of giraffes and Paris could occur by random events per se? The deeper question is, presuming no fraud or false memory is involved, what does all this mean, both generally and specifically?

Before we address the statistical significance question, it is valuable to view these events in their real-life context, so that the personal (and maybe greater) significance of giraffe–Paris co-occurrences becomes clear.

Briefly, I was contacted by a physician whom I referred to as Dr. G. (a pseudonym) on June 9, 2014. Dr. G. called me on behalf of a successful businessman and dear friend of his whom I referred to as Mr. D. This Mr. D. was deeply grieving the recent death of his parents whom he dearly loved, and he desperately wanted to speak with a skilled research medium (a person who had been scientifically documented to be genuine—that is, was not a professional magician or fraud—and could be trusted).

Having published three books describing compelling evidence of genuine mediumship, documented in carefully scrutinized individuals using single-blind, double-blind, and triple-blind protocols (Schwartz 2002, 2005, 2011), I was in a good position not only to recommend highly responsible and skilled research mediums, but to potentially make the request that these mediums make an exception and provide Mr. D. with quickly scheduled readings.

I was able to speak with Mr. D. on June 14, 2014. Based on this conversation, I decided that I should try to pull out all the stops and see if I could arrange for Mr. D. to have long-distance phone readings with three gifted research mediums. Mr. D. is an open-minded yet skeptical person, and he is deeply interested in research in this area. We agreed that if I could arrange for him to have private readings with three different mediums under blinded conditions, and if the readings were especially evidential, that these experiences might help him and his family with the grieving process.

Unbeknownst to me, Mr. D. decided to perform a secret experiment and see if any of the blinded mediums could bring through information regarding two of his mother's favorite things—giraffes and Paris! I later learned that two of the mediums provided highly evidential information about his mother and father, including their personalities, the circumstances of their deaths, and the like; however, none of the mediums specifically brought up giraffes or Paris per se.

As it so happened, I was scheduled to be in New York City on June 25 and 26 to serve as a Visiting Scholar in Dr. Lisa Miller's Spirituality Mind Body Institute, Summer Intensive Master's Degree Program, at Columbia University. I was invited to spend a day and a half teaching the students about postmaterialist psychology and the emerging convergence of science and spirituality. This included theory and research concerning the existence of the living soul (for example, contemporary mediumship research) and the existence of a greater spiritual reality (for example, contemporary synchronicity research as related to the God Hypothesis; Schwartz 2006).

I decided to feature a discussion about God, synchronicity, and postmaterialist psychology for these students, especially since my initial article (Schwartz 2014) reported synchronicities that included Dr. Miller[1], the director of the summer institute (and then-coeditor of the journal).

Meanwhile, it so happened that both Mr. D. and Dr. G. lived part of the time in New York City. In fact, it was possible for me to have the opportunity to meet with both Mr. D. and Dr. G. on June 27 following Mr. D's readings. However, prior to coming to New York, I had no idea that an extraordinary combination of 15 synchronicities involving giraffe–Paris co-occurrences was about to unfold, which connected the three of us, and that this collection of spiritually transformative synchronicities would serve as an ideal exemplar of real-life evidence of spirituality in practice for the Master's students at Columbia.

Now that you can appreciate the collective significance of the emerging 15 giraffe–Paris co-occurrences for Mr. D., Dr. G., the Columbia students, and me, we can address the statistical probability of the 15 giraffe–Paris co-

1 Coincidentally, Dr. Miller gave the opening Keynote Address at the Third International Conference on Science and God, and I gave the ending presentation at the conference. My decision to use the giraffe–Paris example was partly in appreciation of Dr. Miller.

occurrences happening by chance. As detailed in the article, using the super conservative value of a given giraffe–Paris co-occurrence happening 1 in 100 times by chance, the probability of getting 15 of them in a short period of time by chance is 100^{15} or 1 followed by 30 zeros.

What does such a large number mean?

Consider the following. Let us say that it takes one minute for us to observe a giraffe event (reference, object, or picture) followed by a Paris event (reference, object, or picture). Let us also say that we will actively search for pairs of giraffe–Paris events every minute, 24 hours a day, every day, 365 days a year, for as long as it takes for 15 such co-occurrences to occur by chance in a brief time span (that is, 12 days).

How long, in total years, would this take? The answer is astronomical.

The age of the universe is generally believed to be 13.7 billion years. There are 525,600 minutes in a year (60 minutes in an hour, multiplied by 24 hours in a day, multiplied by 365 days in a year). The total number of minutes in the age of the universe comes out to be 7,200,720,000,000,000, which is approximately 72 followed by 14 zeros. Meanwhile, the conditional probability estimated for the 15 giraffe–Paris pairs was 1 followed by 30 zeros. What is the takeaway message? *It would take much longer than the age of the entire universe, by orders of magnitude, for such a super synchronicity sequence to be observed by chance.*

I should acknowledge that the published conditional probability calculation was oversimplified, and that it did not adjust for the fact that giraffes and Paris, for example, have existed on Earth for a tiny fraction of the actual age of the physical universe. However, even if the conditional probability was made substantially more conservative and reduced by a factor of billions or trillions, the final value would still be much longer than the age of the universe.

Importantly, this is just one example of more than a hundred super synchronicities that I have documented since I witnessed my first super synchronicity in the 1980s when I was a professor of psychology and psychiatry at Yale University. Interestingly, I witnessed the most recent super synchronicity category (yellow car events) in the middle of February through

Step 12 Divine Mediation – Collective
Step 11 Divine Mediation – Personal
Step 10 Higher Spirit Mediation
Step 9 Spirit Mediation
Step 8 Human Intention and Energy Mediation
Step 7 Biophysical and Social Self-Organization
Step 6 Geophysical and Astrophysical Forces
Step 5 Psychological and Social Causes
Step 4 Beyond Probable Chance
Step 3 Chance Coincidences
Step 2 Selective Attention (including expectation and salience)
Step 1 Deception – Self (misinformation) or Other (disinformation)

Figure 1 A hierarchy of possible explanations for super synchronicities.

early March 2022, and at the time I wrote this paper, the number of those sequential events had reached 51.

Staircase of Possible Explanations for Super Synchronicities

It is possible to organize possible explanations for the existence of super synchronicities in terms of a hierarchy of explanations, from the simplest and most skeptical steps (Steps 1 through 3 shown in **Figure 1**) to the most sophisticated and most spiritual steps (Steps 9 through 12 in **Figure 1**).

Time does not permit us to analyze and interpret the viability of each of these hypotheses (see Schwartz 2017). However, what should be clear is that given the extreme improbability, as well as extraordinary complexity, of the exemplary giraffe–Paris super synchronicity discussed in this presentation, the justification for seriously considering the Step 11 and Step 12 super intelligent Mind explanations, and therefore the active pursuit of Source Science, should be self-evident. Though such evidence is often mind-boggling, the fact remains that the existence of real-life super synchronicities implies the active and intelligent participation of a Source Mind in everyday life.

Implications of Greater Reality Living for the Earth's Future?

The implications of emerging Source Science for human growth and transformation, and by extension, our evolved co-stewardship of the Earth, expands our perspective on the process of guidance, organization, and design in all systems at all levels, be they scientific and technological, educational and clinical, personal and professional, political and societal, legal and spiritual, or environmental and global. The idea that we live in an integrative and holistic greater reality, and that we co-create with this greater reality, whether we are aware of it or not, touches all aspects of life.

Here is how I expressed this vision in the Afterword in *The Sacred Promise* (Schwartz 2011) book, updated for this presentation. The section was titled "If not now, when? If not us, who?" Note that in this context, the word "Spirit" includes Source plus other levels of Spirit beings, including human souls.

> *If Spirit is not only with us but is calling on us to wake up and see the light, then if ever there was a time for us to get to work, this is it.*
>
> *If you consider the essence, the big picture, of what you have read in this book to be true—recognizing that frontier scientific methods and findings are out there—then there is the serious possibility, if not probability, that spirits and Source are awaiting our active collaboration with them.*
>
> *They are here with you and me right now, awaiting our awakening.*
>
> *Not only that, but they are claiming to be at least as dedicated, trustworthy, genuine, and capable of helping us save the planet as the most highly evolved, responsible, and caring individuals living in the physical world today.*
>
> *They are calling us with a sacred pledge to be with us, all the way.*
> *This is the Sacred Promise.*

The totality of the evidence supports the thesis that the opportunity for humanity to receive guidance, organization, and design information (and inspiration) from the greater reality is there, with the proviso that the use of this knowledge is in the service of universal love and the caring co-stewardship of the planet we have been gifted to enjoy.

It has been said, "Our life is a gift from God; the way we live our life is our gift to God." Can a Source Science inspired, evidence-based approach to life, potentially decrease humanity's inhumanity to itself (for example, from physical and sexual abuse to local and global wars) and to its environment—that is, to nature and the Earth?

Yes, the question of accuracy and integrity of receiving such information needs to be addressed, but scientific and technological efforts are underway to address this fundamental challenge (for example, see https://www.thesoulphonefoundation.org/).

The recently published academic book, edited by Thomas Plante and me, titled, *Human Interaction with the Divine, the Sacred and the Deceased: Psychological, Scientific, and Theological Perspectives* (Plante and Schwartz 2021), places this opportunity in a historical as well as contemporary scientific context. In addition, practical suggestions regarding "greater reality living" for adults (Pitstick and Schwartz 2018), and "the big picture of life" for children (Mapes, Pitstick, and Schwartz 2021), point to the evidence-based potential for human health, happiness, and harmony with nature and the Earth.

If any single twenty-first century scientist expresses this frontier science and spirituality vision, it is Dr. William A. Tiller and his colleagues (see https://www.tillerfoundation.org/), to whom this presentation is dedicated. Finally, hope for a healthy and happier planet comes from visionary private companies, bravely bridging contemporary frontier science and technology with ancient as well as contemporary understandings, as they mindfully employ an evidence-based, interdimensional, co-creative and directive paradigm, such as Focused Light-Force Energy (see www.flfe.com). The International Conferences on Science and God provide an ideal environment for nurturing this essential evolution.

Postscript – The Giraffe Meaning and Symbolism: A meaningful coincidence?

After completing the first complete draft of this presentation, I was inspired to look up the meaning and symbolism of the giraffe on the Internet, understanding the limitations in assuming that such information is either accurate or representative. I went to a well-known site for such information (https://www.spiritanimal.info/) and looked for its entry on the giraffe. The information seemed curiously and coincidentally relevant, and I have boldfaced some especially meaningful phrases from the site.

> *Not many people may think of giraffes in terms of spirit animals, but given their gentle nature and long-necked height they are an ideal representation for a perfect spirit guide or power animal.* **The giraffe's long neck allows it to reach heights above others.** *It has a lot to offer us as* **an inspiration when it comes to understanding our nature and how to relate to others.** *Giraffes have no trouble at all when it comes to* **feeling empathy for others, leading you to non-violence, to sincere feelings, to thought-through decisions.** *The symbolism associated with this animal is centered around the* **themes of peace, the heart, the ability to put life into perspective, and even the development of intuition.**

As for the co-occurrence of the giraffe with Paris, in the environmental world, Paris is associated with The Climate Agreement signed by 195 nations to reduce greenhouse gas emissions (also a curious coincidence). Its relevance to ICSG III is self-evident. ◊

Dedication

This presentation is dedicated to William A. Tiller, PhD (1929-2022), a mainstream material scientist and visionary "psychoenergetic" scientist whose professional and personal life model the best of science and spirituality.

References

Dossey, L. 2013. *One Mind: How Our Individual Mind is Part of a Greater Consciousness and Why It Matters.* Carlsbad, CA: Hay House.

Harris, E. "Giraffe Spirit Animal." *Spirit Animal.* Accessed on July 14, 2022. https://www.spiritanimal.info/giraffe-meaning/

Laszlo, E. 2007. *Science and the Akashic Field: An Integral Theory of Everything.* 2nd Edition. Rochester, VT: Inner Traditions.

Mapes, K., M. Pitstick, and G.E. Schwartz. 2021. *The Big Picture of Life.* Cardiff, CA: Waterside Productions.

Matzke, D., and W.A. Tiller. 2020. *Deep Reality: Why Source Science May Be the Key to Understanding Human Potential.* Cardiff, CA: Waterside Productions.

Pitstick, M.R., and G.E. Schwartz. 2018. *Greater Reality Living: Integrating the Evidence for Eternal Consciousness into Your Daily Life.* 2nd Edition. CreateSpace Independent Publishing Platform.

Plante, T.G., and G.E. Schwartz. 2021. *Human Interaction with the Divine, the Sacred and the Deceased: Psychological, Scientific, and Theological Perspectives.* England, UK: Routledge.

Portela, L. 2021. *The Science of Spirit: Parapsychology, Enlightenment and Evolution.* Jefferson, NC: Toplight

Schwartz, G.E. 2002. *The Afterlife Experiments: Breakthrough Scientific Evidence of Life After Death.* New York, NY: Atria Books.

———. 2005. *The Truth about Medium: Extraordinary Experiments with the Real Allison DuBois of NBC's Medium and other Remarkable Psychics.* Charlottesville, VA: Hampton Roads Publishing Company, Inc.

———. 2006. *The G.O.D. Experiments: How Science is Discovering God in Everything, Including Us.* New York, NY: Atria Books.

———. 2011. *The Sacred Promise: How Science is Discovering Spirit's Collaboration with Us in Our Daily Lives.* USA: Atria Books/Beyond Words.

———. 2014. "God, synchronicity, and postmaterialist psychology: Proof-of-concept real-life evidence." *Spirituality in Clinical Practice* 1 (2): 153–162. https://psycnet.apa.org/doi/10.1037/scp0000017

———. 2015a. "God, synchronicity, and postmaterialist psychology II: Replication and extension of real-life evidence." *Spirituality in Clinical Practice* 2 (1): 86–95. https://psycnet.apa.org/doi/10.1037/scp0000038

———. 2015b. "God, synchronicity, and postmaterialist psychology III: Additional real-life evidence and the higher power healing hypothesis." *Spirituality in Clinical Practice* 2 (4): 289–302. https://psycnet.apa.org/doi/10.1037/scp0000079

———. 2017. *Super Synchronicity: Where Science and Spirit Meet.* Waterfront Digital Press.

Schwartz, S.A., M. Woollacott, and G.E. Schwartz, (eds). 2020. *Is Consciousness Primary?* Battle Ground, WA: AAPS Press.

Session 3 | First Commentary
on the Presentation by Professor Gary Schwartz

A Super Mind, Planetary Well-being, and the Complex Global Crisis: Emergence of the Second Generation of a Global Civilization

Dr. Kazuo Takahashi
Director, DEVNET International, Japan

1. A Super Mind

On the eve of the emergence of the second generation of a global civilization, Professor Gary E. Schwartz has made a very timely presentation. He pointed out that:

The totality of the evidence supports the thesis that the opportunity for humanity to receive guidance, organization, and design information (and inspiration) from the greater reality is there, with the proviso that

About the Speaker

- Director, DEVNET International, Japan
- Former Professor, International Christian University, Tokyo
- Trustee and Advisor, Earth Charter, Japan
- Committee for Promoting Global Governance, Japan's Diet; Co-Chair of Environment Issues
- Worked for the Organization for Economic Cooperation and Development, and the Sasakawa Peace Foundation
- PhD in Political Science, Columbia University

the use of this knowledge is in the service of universal love and the caring of co-stewardship of the planet we have been gifted to enjoy.

Having just entered the revolutionary period of academic research activities with easy accessibility to supercomputers, this "evidence" will be amply substantiated in the coming period. In the fields of social sciences and humanities, the importance of traditional disciplinary lines will decline, and the emphasis of research activities will be shifted to issues and problems, including a search for a Super Mind.

2. Planetary Well-being

Dr. Schwartz's thesis has a number of important implications for the global community. I would like to point out just one of them, which was presented by policy leaders as a legacy of the twentieth century, namely, the Earth Charter. Its preamble starts as follows:

> We stand at a *critical moment in Earth's history, a time when humanity must choose its future. As the world becomes increasingly interdependent and fragile, the future at once holds great peril and great promise. To move forward we must recognize that in the midst of a magnificent diversity of cultures and life forms we are one human family and one Earth community with a common destiny.... Toward this end, it is imperative that we, the peoples of Earth, declare our responsibility to one another, to the greater community of life, and to future generations.*

Toward the end, the preamble states:

> *The spirit of human solidarity and kinship with all life is strengthened when we live with reverence for the mystery of being, gratitude for the gift of life, and humility regarding the human place in nature.*

Having been announced in June 2000 in the Hague, Netherlands, it provided powerful moral support to the Millennium Declaration, which was adopted by the United Nations Summit in September 2000. In the Johannesburg Conference commemorating the tenth anniversary of the Rio Summit on the environment and development in 2002, ESD (Education for Sustainable Development) was adopted, inspired by the Earth Charter. In 2015, SDGs (Sustainable Development Goals), which were modeled after the Earth Charter, were adopted by the United Nations General Assembly.

Based on these records, a new project will be announced in June 2022, at the Stockholm+50 meeting, commemorating the Fiftieth Anniversary of the United Nations Conference on the Human Environment. The concept of planetary well-being will be translated into indexation around the four pillars of the Earth Charter, those being: 1) respect Earth and life in all its diversity; 2) care for the community of life with understanding, compassion and love; 3) build democratic societies that are just, participatory, sustainable, and peaceful; and 4) secure Earth's bounty and beauty for present and future generations. Contributions of each nation (over 200) to the planetary well-being will be measured and ranked each year. The indexation will be improved over time and is aimed to provide the world community with an alternative or supplement to GDP, expecting a fundamental and durable change in our perception of activities.

3. The Complex Crisis

It is important to recognize the historical context in which Dr. Schwartz's statement and the launch of the planetary well-being index are taking place. It is the complex crisis of the world community that has been brought about by the combination of the COVID-19 pandemic, Putin's war of aggression in Ukraine, and the imminent crisis of the environment, starting with climate change.

While humanity has been continually suffering from epidemics, drastic societal changes have taken place when they are combined with other grave challenges. The combination between pandemics and wars have brought about complex crises, leading to civilizational changes. The Peloponnesian war, which ended with the epidemic in Athens in the fifth century BC, marked the end of the classic Greek civilization based on city states, eventually opening the era of empires. The Hundred Years' War combined with the pandemic of the fourteenth century brought about the end of the feudal period in Europe, leading to the Renaissance period. World War I combined with the Spanish flu, which killed more people than the war itself, undermined the spirit of the Europeans so seriously that Pax Britannica was eventually shifted to Pax Americana.

Today, the combination of not only the pandemic with the war, including the potential use of atomic bombs, but also with the imminent environmental crisis is strongly felt in the world community. This clearly points to a historical moment of civilizational change.

4. The Emergence of the Second Generation of a Global Civilization

Over the past several centuries, the European regional civilization has developed into a global civilization. Although producing many vices and atrocities, it has articulated a number of important concepts, institutions, and practices, including human rights, democracy, rule of law, market-based economy, civil society, and science and technology. However, the limits of the first global civilization are clearly revealed by the current complex global crisis, and we are at the threshold of a second global civilization. By elaborating the assets of the first global civilization further, it is our task to articulate the concept, institutions, and practices of the new civilization in a manner that integrates the wisdom, perceptions, and customs that have been developed in Asia, the Middle East, and other parts of the world over thousands of years. Dr. Schwartz's Super Mind is an encouraging contribution to these efforts of humankind by implying the potential harmony of the emerging civilization with a proper mindset on the part of those who promote the new civilization. The Earth Charter ends with the following sentence:

Let ours be a time remembered for the awakening of a new reverence for life, the firm resolve to achieve sustainability, quickening of the struggle for justice and peace, and the joyful celebration of life. ◊

References:

Earth Charter. "The Earth Charter." *Earth Charter.* Accessed on July 13, 2022. https://earthcharter.org/read-the-earth-charter/preamble/

Session 3 | Second Commentary on the Presentation by Professor Gary Schwartz

A Vision for Earth's Future Arising from Frontier Science

Dr. Beverly Rubik
President and Founder, Institute for Frontier Science, Oakland, California, USA

Dr. Gary Schwartz (2022) focuses on fascinating phenomena called "super synchronicities," which he considers among the best evidence for unseen dimensions of reality that point to an underlying intelligent design. He defines super synchronicities as patterns of six or more serial coincidences in our lives that cannot be explained in simplistic ways.

Here is a brief background. Before science, the prevailing cultural myth was a Divine order of oneness of the cosmos, nature, and humanity. This was shattered by modern science with its worldview of mechanistic reductionism. Eventually, even people came to be seen as separate and distinct objects, connected to each other only through direct contact. Subjectivity was dismissed

About the Speaker

- Founder and President, Institute for Frontier Science, California, USA
- Editorial board member of the *Journal of Integrative and Complementary Medicine, Evidence-Based Integrative Medicine, Integrative Medicine Insights,* and *Journal of Vortex Science and Technology*
- PhD in Biophysics, University of California, Berkeley, USA

as irrelevant. Human nature became truncated, and our intimate bond with nature was severed.

Despite this, synchronicity can have a profound effect on our sense of connection to Source and nature. It is a notable aspect of Jungian psychology, of acausal connection, nonspatial and nontemporal. Carl Jung (1973) wrote that synchronicity reveals facets of the unus mundi, the underlying unity of the world. Jung's viewpoint recovered the meaningful connections between the subjective and objective realms.

Super synchronicities can be extraordinary experiences. They can evoke felt meaning, sense of purpose, heightened emotions, and even personal transformation. Those who experience super synchronicities may feel wonder, delight, amazement, greater spiritual connection, and humility in light of a Divine order and creativity. They may also feel love, joy, and gratitude.

Thus, synchronicities can trigger expanded psychospiritual states of consciousness. Such expanded states, including a profound experience of our interconnection with the cosmos, could help bootstrap humanity to recover a lost relationship with nature and the environment. Persons in this state also frequently experience even more synchronicities, which further reinforce their heightened state.

Other people may feel cut off from the greater whole. To them, a synchronicity appears to be a rare and random incident, rather than a meaningful aspect of temporal order and creative unfolding of the universe. These people are in a contracted state. When one feels separate from nature, then nature may be seen as mainly resources to be consumed. Pollution of the environment becomes rampant. A sense of separation from nature, along with a lack of awareness that nature nurtures life, are at the root of environmental degradation.

Dr. Joo's invitation letter indicated that this conference is to "discuss the frontiers of science and explore innovative technologies they offer to solve environmental problems." Frontier science has revealed some new discoveries about emotions that could lead to some helpful technology. Therefore, I would like to briefly mention our work at the Institute for Frontier Science.

One of our main projects is research on the human biofield or energy field (Rubik 2002, 2015; Rubik et al. 2015). We can measure the extremely low-level light emission from the body, also known as biophotons. One subject, an energy healer, intentionally moved from a neutral state to sending "healing love," and we then measured almost 4 times more biophotons near the heart (Rubik and Jabs 2017). In experiments with several human subjects, we found that biophoton emission from the body can change according to intent, emotion, and state of consciousness. The biofield may be seen as a bridge between consciousness and the physical body.

Besides biophotons, we are investigating other components of the biofield. This involves developing novel detectors. We found a coupling between the dynamic emotions experienced over time by a subject and certain physical detectors that were shielded from conventional energies. In other words, the parameters of certain physical systems changed in a predictable way with the human subject's emotions. Positive and negative emotions, respectively, led to changes in opposite directions of the parameters. We interpret this as a subtle informational field that couples the subjective and objective realms. It is also possible that there is a "subtle energy" associated with emotions. We have a prototype of this detector as part of our "sensor suite," which also includes conventional environmental and physiological detectors (Jabs and Rubik 2019). Our so-called "subtle energy" detector shows the greatest response on a computer data acquisition system when a human subject experiences positive emotions such as love and joy.

Although this technology needs further development, we foresee applications, including biofeedback, to help people engage more in positive emotions, which are also important for optimal health and well-being. Such technology might help boost the time spent in heightened psychospiritual states and increase emotional intelligence.

In today's world, many people interact more with communication technology and less with real people or nature. Cultivating love, joy, and gratitude counteracts the increasingly robotic shaping of humanity by lifeless computer programs in our educational systems and work life.

Positive emotions have also been shown to energize water. Consider that water constitutes approximately 70% of the human body, and 70% of the Earth's surface is covered with water. Previously, water was regarded as only a passive constituent of life, a solvent for biomolecules. However, the frontier science of water demonstrates that water plays an active role. Indeed, it is the matrix of life itself. Using the Gas Discharge Visualization (GDV) Camera Pro, a digital Kirlian camera, we measured greater intensity and larger patterns of light emitted from water droplets when human subjects expressed love and sent positive energy to the water samples (Rubik 2011).

If people would focus their love on water in our environment, I hypothesize that it could have a beneficial impact. With loving kindness and greater awareness that we are an intimate part of the cosmos, humanity can choose to live differently—in a nurturing relationship with nature and our environment.

Synchronicities bring about a sense of wonder and an expanded vision of reality. They can infuse our lives with meaning and love, providing a sense of oneness with Divine Source and with the universe. They can uplift us to heightened states of consciousness that can change our relationship with nature and one another.

Pierre Teilhard de Chardin (1959) wrote, "Someday, after mastering the winds, the waves, the tides and gravity, we shall harness for God the energies of love, and then, for a second time in the history of the world, man will have discovered fire."

Reaching heightened states of being, elucidated by a French scientist and theologian—is this one possible interpretation of Dr. Schwartz's "giraffe and Paris" synchronicity? ◊

References

Jabs, H., and B. Rubik. 2019. "Detecting subtle energies with a physical sensor array." *Cosmos and History: Journal of Natural and Social Philosophy* 15 (1): 171–192. Accessed on July 13, 2022. https://cosmosandhistory.org/index.php/journal/article/viewFile/782/1405

Jung, C.G. 1973. *Synchronicity: An Acausal Connecting Principle.* Princeton, NJ: Princeton University Press.

Rubik, B. 2002. "The Biofield Hypothesis: Its Biophysical Basis and Role in Medicine." *Journal of Alternative and Complementary Medicine* 8 (6): 703–718. https://doi.org/10.1089/10755530260511711

———. 2011. "The power of 'magnetized' water." In *Spiritism and Mental Health,* edited by E. Bradgon. Philadelphia, PA: Jessica Kingsley Publishers, 201–211.

———. 2015. "The biofield: bridge between mind and body." *Cosmos and History: Journal of Natural and Social Philosophy* 11 (2): 83–95. Accessed on July 13, 2022. https://www.spiritualtechnologies.io/wp-content/uploads/2016/01/Rubik-Biofield-bridge-between-mind-and-body-Cosmos-History-2015.pdf

———, and H. Jabs. 2017. "Effects of intention, energy healing, and mind-body states on biophoton emission." *Cosmos and History: Journal of Natural and Social Philosophy* 13 (2): 227–247. Accessed on July 13, 2022. https://web.archive.org/web/20180423014629/http://www.cosmosandhistory.org/index.php/journal/article/viewFile/608/1027

———, D. Muesham, R. Hammerschlag, and S. Jain. 2015. "Biofield Science and Healing: History, Terminology, and Concepts." *Global Advances in Health and Medicine* 4 (Suppl.): 8–14. http://www.gahmj.com/doi/abs/10.7453/gahmj.2015.038.suppl

Schwartz, G.E. 2022. "A Vision for Earth's Future Arising from Frontier Science." Invited presentation, Third International Conference on Science and God (ICSG III), "Environmental Restoration in the Era of Frontier Science." Washington and Seoul: HJIFUS Press.

Teilhard de Chardin, P. 1959. *The Phenomenon of Man.* New York: Harper & Row.

Session 3

Discussion

GunWoong Bahng (Session Chair): Thank you very much for the nice presentations and comments. Now we will have the question-and-answer part of our session. If you have any questions on the topics presented by the speakers or commentators, just raise your hand and ask the question. Please specify to whom you are addressing the question and what you are asking about. Were the topics so clear that no one has questions? Lee Jae Young is here with a question. Please go ahead.

Jae Young Lee: I have a question for Dr. Schwartz. In this presentation, you described the transcendental consciousness and super simultaneity. I would like to ask a question while fully agreeing with your idea. Is this transcendental consciousness you mentioned, developed by yourself or is it given to you from another transcendental world?

Gary Schwartz: I am not sure that I understand the question. Could you repeat the question, or could the speaker ask it in a slightly different way?

Lee: Yes, you mentioned transcendental consciousness. Was it developed by yourself, and is it given by other work? Does that make sense?

Schwartz: No, what do you mean by "given by other work?" Is it given by God and any transcendental void? The observations of super synchronicity typically are external events or events in your consciousness, where often you see the patterns and directions, but without necessarily knowing how to interpret those patterns, as well as what the "guidance of the presence" means. To do that requires one to be open to asking for and receiving guidance, both internally and with others who are also open to receiving them, typically the "hidden meaning" in the patterns. One of the questions that people often ask, which is I think a very deep one, is why the universe, the Super Mind, and the Divine present these patterns where we are left to seek their deeper meaning, as opposed to just being given the information.

It is as if we have to be mature enough and care enough to want to seek the deeper significance in order to be able to receive the higher guidance. Yet, there are people I have met from different places all over the world who are actively seeking such wisdom and guidance from a greater reality. This is also where technology is going, and I think it will be available in the future, as Dr. Rubik spoke about. We, too, have been working on developing technology of literally extending the notion of being able to detect these organized energies and higher energies in such a way that active communication, like what research mediums and higher channelers receive. There was some discussion of channeling earlier on in this in the conference.

I am not sure if this helps, except to say that the opportunity for us to grow in these capabilities is on the one hand, nascent, and on the other hand, prescient. I think it is part of the human opportunity. I knew there was a time in human history, for example, before we had musical instruments or technology, especially with this kind of communication. If you grow up in the absence of all this technology, you would never know that humans could invent musical instruments, ultimately create music, or eventually develop capabilities to communicate over long distances.

Ultimately, now in real time, we can communicate around the world and see each other. Technology is opening our awareness, evermore expanding our awareness and the opportunity to continue to develop that so that other beings, including our ancestors who have been on the Earth before, to develop a more active relationship with and co-stewardship over nature. This all the way to the Source is, again, one of those future opportunities.

Bahng: Okay, thank you. The next person with a question is Pierre Beauregard.

Pierre Beauregard: Hello, Prof. Schwartz. It is an honor to have a chance to ask you a question. I really appreciate your talk on super synchronicities and the clear message that there is a Super Intelligence trying to communicate with us and asking for our collaboration. In consideration of the environment, COP25 in Paris, and all of this, my question is the following. Could it be that postmaterialist science is emerging as a new kid on the block, so to speak, on the Earth, where the old wisdoms through religions have been trying to be the ones to guide us? Postmaterialist science is new and can it, as the new kid on the block, communicate with and get help from the Divine to unite all these various trends of thought of how to communicate and receive wisdom, from the wisdom of old? Could postmaterialist science thus emerge as a new religion or approach to bring humanity together and communicate with this Super Mind, or God?

Schwartz: That is a beautiful question. Thank you for asking. In a deep sense, of course, science is the new kid on the block. I mean, science is now rediscovering some very ancient traditions, re-appreciating them, and even resurrecting them in some deep sense. These are concepts that were discarded or treated merely as superstitious, signs of ignorance, or whatever. All of us—both scientists and "non-scientists," including people who are deeply religious—have to go through a process of reeducation. There are these gems of wisdom and truth that go back thousands of years, in all cultures. They have universal commonalities, and we have different languages, come from different perspectives, and have different paths to the same ultimate unity.

The idea is, that is where humility and discernment become so important. My hope is that this is where the merging of science, spirituality, and religion can come to pass, if all of us are willing to say that we have more to learn. That includes looking at things in a new way, which means re-appreciating things we may not have appreciated before, and giving up certain ideas that we just assumed were true. They turn out to be not the way things really are, and that is the history of science. I think all of us need to grow in that area.

Bahng: Okay, thank you. The next question is from Staffan Berg.

Staffan Berg: Thank you very much for this opportunity to ask a question, Dr. Schwartz. Thank you for mentioning Dr. Tiller. I have deep respect for him and fascination. But that is not the question here. I would like to ask you more about the super synchronicity you mentioned. After having recorded 51 events, have you found some qualitative, underlying explanation in addition to just their occurrences? I am sure you are thinking of the what, why, how, or some explanation. Also, if you can touch upon this, you also mentioned the co-creatorship of reality. Have you found that you could actually trigger some of these synchronous events yourself before they happen? Have you seen that?

Schwartz: Yes, thank you. That was a wonderful question, and let me see if I can do justice to it to some of them—my experience after many years of tracking super synchronicities. As an aside, the beauty of tracking synchronicities and super synchronicities is that any of us can do so if we are so inclined, which I call "self-science," by applying sort of the basic methods of scientific seeking to the laboratories of our personal lives. Then we can also develop super synchronicity communities, with people whom we shared events, upon which even deeper constellations of events can occur. Getting back to your question, my sense is that the meanings can operate at different levels. For example, there could be a specific, focused direction for me and my personal life for a given event. However, there also can be a message or lesson about how super synchronicities operate at another level, on a broader or general level. There can be higher level meanings of implications for humanity.

Believe it or not, all of those levels could be present in a given sequence of super synchronicities. It can thus become very complicated very quickly. For example, the notion of complex scripts or narratives in which you have different storylines and messages being told—with we humans being able to create and being neophytes—compared to the implicit stories, narratives, scripts, plans, and descriptions that relate to our lives. If some of you are interested, I talk about some of this in this book on super synchronicity, so you can learn about some of this there. In my most recent synchronicity, with super synchronicity over the past couple of months, there literally was information that pertains to your second question. There were pre-events that happened that were anomalous, which I did not even realize were at the beginning of a super synchronicity set of events, until some more obvious ones happened. It was only after the more obvious ones happened that I realized this was the case.

As I was trying to make sense of why this particular set of events occurred, I realized that other ones had been happening, which I had not seen the connection with, and they had a key to the bigger or deeper spiritual meaning. For the last question, I should also briefly mention that there are things that we can do that will increase the probability of us seeing and receiving synchronicities. I am not sure about how much we are "producing them," causing them to happen more as in inviting them. My experience has been that the more grateful I am, I come to witness and experience synchronicities. And the more I honor them when they occur, the more likely it is that they will occur.

Also, if I make requests to the universe to help me make better sense of the patterns of information that I am receiving—as in the old wisdom, "Ask and you shall receive"—under the right conditions. We could spend hours discussing your question, and I apologize for just briefly touching on that very deep question.

Bahng: Thank you, Dr. Schwartz. We have two more people with questions. Before going further, I think that the audience is very interested in the topic of super synchronicities due to most people having had that kind of experience in their life. Thus, I think that people are asking questions to you. However, I would also like to encourage the audience to ask questions to Chris Laszlo, as he showed a way to increase synchronicity with nature. Let us move on to a question from Takahiro Hiroi.

Takahiro Hiroi: I am sorry, as my question is also for Dr. Schwartz. I appreciate that you are pursuing the soul phone project. As you may know, the co-founder of ICUS, the late Rev. Sun Myung Moon, said we should develop a spirit world communicator so that people can know clearly what happens after death and beyond, making it so that people cannot commit sin. They have to know the consequences of this earthly life. I am sure you might be thinking of that kind of positive effect of your device. Such a device can communicate with the spirit world or dead people, and we ourselves have a spirit, although we are in the physical world. My question is the following. Do you not think that how such a device functions may depend on the spiritual condition of the user?

Schwartz: That is a wonderful and very deep question. There are different predictions, depending upon one's theory. We are attempting to create a technology that is, relatively speaking, independent of the intention or emotional state of the users and is more like a normal phone that we are used to. Whatever that particular technology is, whether the person is happy, sad, angry, frightened, or whether they are grateful or not grateful, the technology can detect their energy and be used for the purpose of communication. That particular theory that guides our work may or may not be correct. If I had a vote for the way nature is, I would actually wish that we could develop a technology where its accuracy would be dependent upon the level of purity, intention, and beneficence of its use. That is one of those wonderful questions for future work. Thank you for asking that question.

Bahng: Thank you. We have the next question from Prof. Masayuki Ohkado. I think we do not have much time, so please ask the question quickly.

Masayuki Ohkado: I am afraid my question is also for Dr. Schwartz, so I apologize. Dr. Schwartz, thank you very much for your very stimulating talk. I was impressed by your research using silicon photomultipliers, detecting the possible existence of the spirit. I then wondered whether you are also thinking that spirits are related to synchronous events and can be detected as you did in your spirit communications research. Also, do you have any other physical, material elements that might be related to the presence of spirits? Those are my two questions. Thank you very much.

Schwartz: You are very welcome. I think we have about two minutes. Regarding the first question about silicon photomultipliers and the emission of light, I have now lost what the actual intent of that question was. Could you run it by me very quickly? Anyone?

Ohkado: Spirits? Related to that?

Schwartz: Oh, yes. The answer to that is, "Yes." When I talk about the hierarchy of possible explanations, from the simplest and most materialistic, up to the highest and most spiritual, I think it is the eighth or ninth level of those 12 levels. It is the role of post-material beings, formerly physical people, using mechanisms that we address in parapsychology, where they can influence our minds or those of animals and occasionally also affect quantum effects in the physical world. They can thus play a role in signs and synchronicities, as well as higher beings. Part of the reason and the primary reason why I focused on the Super Mind and the highest form for this presentation is partly because this is an international conference on science and God. There are not too many organizations that are brave enough to bridge that level. I thus felt it was important that, since super synchronicity speaks to that highest level as well, we include that in the conversation.

Bahng: Thank you very much for the brief explanation. We still have three more people waiting with questions. However, unfortunately, it is time to finish this discussion session. I am sorry that we need to end this question-and-answer session. I very much appreciate all the participants, speakers, and commentators for making this session very valuable, useful, and exciting. This is a good subject. Thank you again, everyone, and I hope to see you through synchronicity someday. ◊

ICSG III

CLOSING SESSION

Environmental Restoration in the Era of Frontier Science

Closing Session | Report on Session 1

Heralding a New Scientific Revolution

Professor Menas C. Kafatos
Session 1 Chair

What a great pleasure to be with you again. I must say that I learned so much yesterday, because of the great panels and great speakers we had. We had two main speakers, or presenters, and for each of the presenters, we had two commentators. I will say a few things about all of them in just a second. The overall theme of Session 1 was, "Heralding a New Scientific Revolution." What does that really mean? I will say a few words about that.

The important point is that, first of all, this dialogue between science and religion is something that is still developing. Although it is more common than it used to be, personally, I believe that it is still not given its true place in the present era of human endeavors and societies. We can speculate about why, but there will be too much to consider. Yet I will try to speculate and give a couple of answers in 15 minutes, given that perhaps it is the nature of the human mind itself that seems to present some obstacles, or apparent obstacles or challenges, that are not easy to overcome.

Science and God, or science and religion—I use the terms God and religion interchangeably—are not really two opposite sides of the spectrum. Rather, they are two sides of the same coin. With that metaphor—two sides of the same coin—we can more easily make our case to the larger communities as to the importance of the dialogue. You know that these days dialogues are very difficult. We have different stripes of international situations; we have things that do not look too good. But here in this forum, all three of us, who were the coordinators, represent a much larger number of people.

While I am speaking for myself, I can practically hear the same message from the other two coordinators and conclude that the message is the same. That humanity is one; that we all need to look towards solutions that unite us rather than divide us. There is a very practical reason for that orientation: solutions that unite us have to do very much with the survival of the species—the survival of many species, not just our human species. Of course, this is the overall theme that we have been following: basically, the ecological, or the health of the ecosystem, or the environmental aspects; and then, going beyond them, the divine aspect, if I may call it that. In each of us, there is divinity that we all represent one way or another, even though we come from different backgrounds and different languages, different schools, different races, different continents. The basic point is that we are all humans.

Then, without further ado, in Session 1, we had two presenters from two great centers.

The majority of the presenters, by the way, were females; perhaps representing the shift in the overall awareness of the Earth towards the feminine aspect. The first presenter was Dr. Marjorie Woollacott, and the second was Dr. Helané Wahbeh.

Dr. Marjorie Woollacott is a Professor Emerita and former chair of the Department of Human Physiology and a member of the Institute of Neuroscience at the University of Oregon. She is an individual who is a deep thinker, easy to work with; but more than anything else, she is a visionary.

Our second presenter was Dr. Helané Wahbeh. She is the director of research at the Institute of Noetic Sciences in Petaluma, California. It is a well-known institute that explores the boundaries beyond the sciences and beyond the *nous,* which is the mind. Noetic science then, means the "mind" science, but it is really more than that.

Together with each presenter, we had, as I mentioned, two commentators. After Dr. Woollacott, the first commentator was Dr. Jan Holden. Jan is a Professor Emerita also, at the counseling program at the University of North Texas, in Denton, Texas. In addition, the second commentator was Dr. Diane Powell. She is a neuro psychiatrist, independent researcher, and treasurer of the Parapsychological Association in Medford, Oregon.

Parapsychology is, in my mind, a forefront field, even though many times my colleagues in science do not think much of parapsychology. I would say that it is in fact, an important area of exploration and is moving forward using the best of psychology, yet going beyond traditional psychology.

After Dr. Wahbeh, the first commentator was Professor Masayuki Ohkado, who is a professor at the Graduate School of Global Humanics at Chubu University in Japan. He also gave us some deep thoughts to contemplate. Finally, Dr. Arnaud Delorme, whom you may know, and I know personally, is a research scientist, also at the Institute of Noetic Sciences in Petaluma, California. Then, these are just some general introductions.

In our session, the main idea or theme that came through was that there are different ways of looking at things, or different ways of knowing. These ways of knowing go beyond particular, narrow disciplines.

Dr. Woollacott, the first presenter, speculated on why there is so much resistance to change. Coming from the point of view of her background in science, she presented some scientific reasons why, in fact, that may be the case. The difficulty has to do with the senses, or with the sensory system. She made the case that perhaps a way to shift the awareness is to first be aware of the boundaries that are presented by the senses.

Of course, the senses are what make us human; but if we are not aware of their limitations, then our senses themselves are the ones that bind us. This play on words is a sort of "quick and dirty" way to paraphrase what she said. She talked in great detail about the structure of the brain; and then she went on to speculate about the mind, and how the mind in some ways "couples" with the brain. Although one may think that the mind is a derivative of the brain, she made the point that it is way more than that. She made strong cases for the things that perhaps have to do with the special aspects that are so important to all of us.

Then, Jan Holden, who was a commentator, talked about the ability of the mind to actually operate beyond the brain; thus, more or less following the same theme. She emphasized that the mind extends beyond, transcending the physical presence and being aware of transpersonal experiences, so to speak. With that, she talked a little bit about some of the theoretical aspects of feedback to the brain, and of higher states of consciousness, so that they may be called "mirror reality." In that way, she reinforced Dr. Woollacott's talk.

Diane Powell talked about the power to realize, in some ways, the full fruits of what you are doing—if you repeat it, bringing it to your attention. Perhaps there is more than this particular individual method. She also brought in some other ideas, some of the Eastern concepts of reincarnation and karma.

She made the effort to present some practical aspect of how these repetitive programs are, like having some tape recording playing over and over again. They can be thus understood, in a way, as a means for us to be called "ambassadors of peace." She sort of emphasized that we must learn to love beyond our own immediate presence. I think that message resonates with all of us.

Moving along, the presenter and commentators for the second topic dealt more with practical aspects, whereas those on the first topic focused more on theoretical aspects. The first topic emphasized the characteristics of the postmaterialist era. That is what Dr. Woollacott talked about: characteristics of postmaterialism—certainly something that is beyond the material world.

The second topic was on phenomena that require shifts in the scientific paradigm. In other words, if we are making the case, theoretically, that there is a shift, then we scientists must actually show some understanding or some need for evidence to qualify further study. Even though we may not have the answers, at least we are aware that there has to be something that needs to be explored further. For example, telepathy, precognition, or the various "clairs," such as clairvoyance.

Dr. Wahbeh talked basically about the aspects of reality beyond the physical realm, that science in the past was not very kind toward. Nowadays, however, I would venture to say that scientists, including physical scientists, including those in fields like my own, do not just ignore them; because we realize that perhaps there is something about the field of parapsychology. There is something about ESP (Extra Sensory Perception), there is something about Psi, and the psychic, and the noetic. Noetic is a more established term—it refers to the mind, but even beyond the mind. This concept is basically the foundation of experience, which is so difficult for science to get its hands around.

She also talked, by the way, about some of the great experiments being done with random number generators, and how the mind affects the results of random number generators. I find it very interesting—as a scientist in quantum physics—the chance for me to hear this work that, of course, I am aware of. However, it seems that there is quite a bit more awareness out there than I knew. She also talked about quantum entanglement—the idea that everything is entangled, which is very, very interesting.

Then Professor Masayuki Ohkado talked about phenomena not covered necessarily by the presenter. He emphasized some other points. He talked about the channeling of a person's deceased relatives—that was very interesting. He spoke of what is called a "psychic reservoir" —that there may be a huge reservoir of awareness. It consists of not just our immediate relatives but also other humans who have passed away, a repository that we can tap into. In a way, we are bringing together past life, life between life, and then life in the womb, right before one's birth. Of course, if you know that, in the womb, there is awareness, then when the baby is ready to come out, there should be memories that can be retained. The speakers brought out evidence that quite often there are children who remember. There were many questions; however, I do not have time to go through all the questions.

Finally, Dr. Arnaud Delorme, the last commentator, talked more about certain aspects of physics, such as symmetry breaking, and again, the interconnectedness of everything. In terms of the equations he presented, there were some deep thought questions that had to do with: Does it have to be this way, or is there another way to do it? How do we know that? What am I saying? These questions are coming from, I would say, a positive point of view. However, they were also challenging. I mean, when we say something, then we need to be able to support it somehow, and not just say words like this or that out of the blue. I do appreciate that, as a scientist, that way of scrutiny is just the way you ought to be.

Next, there was a question about compassionate meditation, and blissful existence, or awareness. All other matters had to do with the use of smartphones. That was quite a wide topic to cover. Actually, Dr. Wahbeh said that, in a way, these technologies sometimes are thought of as negative. However, these technologies open up developing new vistas that were not available before. In that sense, they not only defeat the older scientific paradigm, but they also, if I may call it that, they feed the spiritual paradigm.

In conclusion, we are very lucky to be living in the twenty-first century, with these technologies that we can use—as long as we do not let them misuse us.

With that, I would like to thank again the organizers, this great organization, and all of you for being present. I guess there are a number of us here, and many more listening to us. I thank you so much, and I look forward to seeing you in the future. Goodbye. ◊

Closing Session | Report on Session 2

Applications of Frontier Science in Addressing Environmental Challenges

Dr. Paul Filmore
Session 2 Chair

Mr. President, Conference Organizing Committee, and Delegates:

Welcome to my outline for this summary of Session 2, "Applications of Frontier Science in Addressing Environmental Challenges." I would like to explain that I will be taking Topic Two first, as it flows more naturally to move from the "external environment," which is related to technology and climate change. The title of Topic Two is, "Breakthrough Energy Technologies derived from New Paradigm Science." I will then move on to what I call the "inner environment," associated with health and healing. This is Topic One, "Diverse Modes of Healing Based on Frontier Science." I will then finish with a brief conclusion.

Dr. Thomas Valone, President of Integrity Research Institute in Maryland, focused on the research and development of new energy technologies, propulsion, and bioenergetics, based on both conventional science and new paradigm science. He led us for Topic Two, "Breakthrough Energy Technologies Derived from New Paradigm Science," and outlined the problem associated with global warming of our biosphere using dramatic graphs. He then looked at breakthrough technologies that we can potentially utilize to stop polluting and remove pollution from our biosphere. Such technologies and issues include carbon removal, food production, vertical farming, water harnessing, atmospheric water harvesting, energy, and other topics. Also discussed were clean energy harvesting from undersea turbines, magma, geothermal, thermal fluctuations, using graphene, osmosis, biomass, water evaporation, and the Casimir cavity device approach, to name but a few. Overall, he left us with a positive outlook, that there is plenty of energy, food, and water available if we develop some of these future energy conversion breakthroughs.

Our first commentator on topic two was Dr. Thorsten Ludwig. He took up and expanded on some of Dr. Valone's technical energy harvesting areas, including magnetism and zero-point energy. However, he noted that it was not enough to change technology, unless we also change our attitude to include peace and love as an essential element in our technological practice.

Our second commentator, Dr. Patrick Bailey, explained that we need to have a mindset that does not say "not possible" but instead have a mindset that thinks "outside the box." I have found this so essential with my consultancy work. The key here is to have groups undertaking a simple task and then having them banging their heads on the wall and saying "It cannot be done." It is only when they are shown a solution is possible, with a radically new approach, that they will not give up or simply use conventional thinking with future problems. I find using the problem-solving technique, TRIZ, a useful primer in this area.

Dr. Bailey also explained that the real problem that we have at present is that governments, particularly mentioning the U.S. which he knew most about, formally classifies (in other words, hides away) many potential breakthrough areas. He suggested a solution that these potential areas with possible solutions need to be reexamined in terms of our world crisis but within secure areas of corporations or government establishments, where they can maintain careful control over what may be dangerous. Dr. Bailey concluded by challenging the attendees by asking what we are going to do about this today.

At this point, I added a comment on governments classifying innovations by talking about the function of corporations. I gave an example of

Sanyo's zero-detergent washing machine that uses ultrasonics to break up the dirt and uses electrolyzed water to remove the dirt. This machine would have used much less power than conventional washing machines, as the clothes only needed a gentle agitation along with far less water due to the lack of a detergent to wash away, with multiple rinses. Also, note that the conventional washing machine pollutes water with detergent, which then we have to remove from our biosphere. So, where is this home appliance now? Were the patents bought up by a large multinational corporation that makes millions on selling detergent? There is no doubt that responding to new ways of doing things will, and do, challenge conventional corporations by, for instance, removing markets; in this case the detergent market. Companies will have to be ethical and far more agile if we are to survive the next few decades.

I will now move on to Topic One, "Diverse Modes of Healing Based on Frontier Science," for which Dr. Christina Ross was our main speaker. Dr. Ross is a researcher and board-certified polarity practitioner from the Institute of Regenerative Medicine, Wake Forest School of Medicine in North Carolina. Dr. Ross introduced us to energy medicine, a branch of integrative medicine that studies the therapeutic effects of subtle energies. She became interested in this from practicing polarity therapy, where this model says that flowing energy is associated with wellness, while blocked energy flow is associated with unwellness. She then gave us a detailed account of her PhD research, which is well worth listening to, but I will not repeat it in detail here.

In summary, this research used Helmholtz coils to produce an aligning field to heal cells under lab conditions. She found the most beneficial frequency to use is around five Hertz. This mechanism of using external pulse EM fields, she suggests, is the same as that used by healers using their hands, to, for example, stimulate bone regeneration or other forms of healing. Dr. Ross has been using pulse EM fields on some patients with severe urinary problems and incredible pain, to extraordinary effect. She hopes for another grant to extend her work and we wish her well there. Dr. Ross discussed a mechanism of how this subtle energy might interact with cells to produce harmonic and coherent enhanced energy interactions that lead to this healing outcome. She finished by discussing how this mechanism of generating coherent fields may also be used by healers, or by any of us, using our conscious directed thoughts to produce healing effects.

Dr. Hiroshi Ishii was our first commentator for this topic. He continued the discussion of resonant phenomena in the field of quantum dynamics by referring to oriental medicine understanding. He gave an example of how an imbalance in the three elements: *Qi* (which is the fundamental energy of life), water, and blood, give rise to illness. He also discussed other mechanism models, which are similar in their understanding of subtle energy, for instance, the human chakra system, or *Qi*. He then discussed the effect on the Western world of George Lucas and his Star Wars films, and their often-repeated concept of "life energy pervading the universe," and that, like the Jedi Knights, we can master this with training. "May the Force be with you" is a phrase often stated in the films. Perhaps the problem here in the materialistic West is that we unconsciously relate to this phrase, but we do not know how to challenge the scientific medical community to practice this postmaterialist science understanding.

Our second commentator, Dr. Won H. Kim, took the theme further about how we in the West perceive medicine as a branch of materialistic thinking, rather than including a mental model of subtle energies. He started with "water memory" and how homeopathy, which has been around for many years, is still not explained by Western scientific research. He suggests that to compensate for this limitation in Western thinking, we need to expand our model by introducing an "imaginary space" component, which perhaps may help broaden the Western "physical only" space component model. That may give an opportunity to open up this closed Western mindset.

Briefly, in conclusion, we see from Dr. Valone's talk that in energy, food, water, biosphere, security, all these areas are possible and achievable. We also see that the health and wellness area from Dr. Ross's talk can be enhanced, probably dramatically, if we allow ways for incorporating energy medicine understanding.

What has become clear from both of these talks, along with the informative commentator remarks, is that paradigm shifts, in other words, changes in mental models on the personal, corporate, and government levels, are essential if we want to continue without cataclysmic change as a species. It is no use anymore, hoping that things will work out by themselves. Instead, we all need to direct our *Qi*, even if it is on a small individual scale, to make changes happen now.

I wish to thank the president and the conference organizing committee for their organizational work, and for taking on the challenge of such an important conference theme. I wish to also thank our excellent session speakers and commentators. Finally, I hope that you, the delegates, agree with me, and are moved to action from what you have heard over the last two days. This action is collectively essential, in both practice and love.

Thank you. ◊

Closing Session | Report on Session 3

Frontier Science Perspectives on Our Relationship with the Natural World

Professor GunWoong Bahng
Session 3 Chair

I will give a brief report of Session 3. It seems that we human beings are losing our connections with nature and the universe, and human relationships are seriously deteriorating. All over the world, people are in miserable states. In this sense, this conference is very timely and meaningful, considering the current chaos and crisis on Earth and in human life. How can we recover our peaceful life on Earth? The main focus of this conference is on this theme, in my understanding.

To understand the theme of this conference more clearly, I would like to start from Session 1. The first session was focused on the new scientific revolution, based on anomalous phenomena such as energy healing, channeling, and near-death experiences, which have been treated as taboo by mainstream science. Because such phenomena are regarded as outside the scope of their research and study range, the revolution cannot take place from within materialistic science.

Actually, materialistic science has greatly contributed to our understanding of nature. However, it has failed to discover a realistic vision and the meaning of existence and life. This is why we are seeking something that goes further, far beyond the materialistic perspective. We expect that a paradigm shift will occur when we understand those so-called "extraordinary phenomena."

The new paradigm was described by new terminology, namely, a "post-materialist paradigm," as coined by Dr. Woollacott. I am sure that those diverse anomalous phenomena, which were discussed by Dr. Wahbeh, will be regarded as normal phenomena in a short time, when we overcome the limitations of the materialistic perspective on Earth and nature.

Following the first session, the topics were expanded to search for the possibility of finding practical applications and technologies, including medical treatments and new free energy sources, in the second session. Presentation topics in this session included diverse candidates for new and effective methods in medicine and energy sources.

These possibilities clearly indicate that there can be a revolutionary approach, which was discussed during the first session for environmental protection, once we change our perception of these technologies. It will provide much more effective and fundamental solutions, compared to the current passive approach to environmental protection, such as carbon capture or carbon-free strategies.

In the third session, our relationship with the natural world and a vision for the future were discussed, in which the relationships are recovered through the approaches that were proposed in the first and second sessions.

The new paradigm in science clearly indicates that we are one and interconnected, even including God. In other words, nature, the universe, and all of life, including human beings, are not separated. We cannot find any boundary lines in the universe. Nature is based on a dynamic living system. If there is no flow of energy, it will disappear, rather than being in a motionless state. Time and space cannot be distinguished from each other. This can be understood quite easily when we think about the freeze-frame of a movie or video. In this state, time does not flow. There is no change.

If space is frozen, what will happen? Frozen space means that the distance between objects does not change. As a result, there will not be any change even as time flows. This leads to the same state as the freeze-frame. Space can be frozen, as time can be frozen. Therefore, everything in nature can exist only when time and space change together. This fact provides the fundamental perspective that nature is in a dynamic state. Nature exists because it dances. We may call time and space itself a kind of movie screen. Here, the screen itself is changing, rather than the illusory media projected on the screen. Now, the next question will be, "Where does the energy come from to make the screen dance?"

An energy source cannot be found in the same dimension because there would be no driving force for an energy flow. This is because there would be no potential difference in the same dimension. There should be an energy source from other dimensions or upper dimensions that has a higher potential to make everything in our dimension alive and exist. The energy flow from other, higher dimensions may be described as the divine love coming from God. In this sense, the upper dimension of time-space can also be a source of free energy. This science, based on the postmaterialist paradigm, may be called the science of the Source.

Thus, recovering our connection with nature and God is essential for the smooth flow of energy and vital for life. Prof. Chris Laszlo described a simple and effective way to recover these relationships. He showed that not only recovering these relationships but also realizing them even in the management of daily life is important and beneficial. The commentators, Dr. Storberg-Walker and Dr. Glen Rein, supported the importance of recovering oneness from their diverse forms of research.

Following this presentation, Prof. Gary Schwartz discussed synchronicity with an awesome example. The example shows that synchronicity occurred even for a case with an extremely low probability. Synchronicity may be understood as a kind of phenomenon that happens when we recover connectedness with nature and God. When we recover the connection with nature and God, Prof. Gary Schwartz predicted that we will be in the divine love of God and have a bright future. Dr. Rubik and Dr. Takahashi, commentators on Prof. Schwartz's presentation, strongly supported this idea, based on their lifelong experience and research results.

In summary, when we fully understand the oneness of the universe and recover the connection with nature and God, the current problems on Earth can be overcome in the right way. As we work together in harmony, aiming for a bright future, we will be in a synchronized status with nature, with the strong feeling that we are one.

This is a brief summary of the third session. Thank you very much for the presentations and active discussion. I also appreciate the hard work of the staff to make this conference successful. Thank you very much. ◊

Closing Session

Closing Remarks

Dr. Douglas D.M. Joo
Chair, ICSG III

Distinguished scholars, scientists and guests from all over the world:

As we have come to the end of the Third International Conference on Science and God (ICSG III), I would like to express my profound gratitude to all of you—session chairs, speakers, commentators, and general participants, as well as the iPeaceTV team—for your tremendous investment that made this a very productive and successful conference, spanning all time zones around the world. I also wish to acknowledge and thank The Washington Times Foundation and *The Segye Times* for their support.

This conference has widened and deepened our awareness of your ongoing valuable efforts to explore and research the future of science in a uniquely nontraditional way. I would like to encourage you to continue this extraordinary work for the benefit of a world that needs unconventional approaches to deal with its existential crises, especially the sources and consequences of environmental problems.

Two years ago, at ICSG I, we overviewed the history of science and recognized that we are entering a new era of "frontier science," which includes a paradigm shift from a materialist to a postmaterialist perspective. Last year, through ICSG II, we saw evidence supporting the view that we are at the threshold of another scientific revolution. A major feature of this revolution is the concept that consciousness, and not matter, is primary in the universe.

Now, at the close of ICSG III, we can confidently say that the postmaterialist era has arrived. Some of the key points that caught my attention at this conference are as follows. First, postmaterialism is not only related to the study of consciousness but also extends to spirituality, affirming that we are spiritual beings. Second, we are all interconnected with one another and with our natural environment. Thus, if we harm others or the environment, we are harming ourselves. Also, phenomena that are called anomalous because they cannot be explained by a materialist philosophy are actually quite common in everyday life. The conference also covered amazing discoveries in frontier science that are helping us with healing illnesses and novel technologies that can produce energy without releasing greenhouse gases.

We have come to recognize that scientific research has its limits, and science based on a materialist philosophy is not enough to protect and preserve the environment. Rather, by treating nature as sacred and expressing divine love toward all things, we can ensure a bright future for our planet and all its inhabitants.

As we have witnessed during the sessions, you have already made tremendous progress toward the change to a postmaterialist worldview. Nevertheless, the shift to a nondual world requires our combined efforts not only on an academic level but also very practically on educational, socio-economic, and political levels. HJIFUS is committed to contribute to this work to develop new research and find new applications and projects that fit this new era, focusing on restoring our natural environment.

Over the coming months, we will gather all the materials and publish the conference proceedings in print and post the speeches on our website. Furthermore, we will introduce the contents through our online magazine, *The Earth & I,* to a worldwide audience.

In closing, I would like to encourage you to fill out the reflection forms and give us your feedback, suggesting timely subjects and expert speakers, and how we can improve our future meetings.

Thank you very much, and God bless you! ◇

ICSG III

Appendix

Organizing Committee

Dr. Douglas D.M. Joo, Chair
President, HJIFUS

Dr. Dinshaw Dadachanji
Research Director, HJIFUS, USA

Mr. Didier Guignard
Founder and President,
Club Science Paris, France

Dr. Chul Hee Han
Professor, Chemistry,
Sun Moon University, Korea

Mr. Nobuyuki Iioka
Deputy Director, ICUS Committee, Japan

Dr. Sung Bae Jin
Chair, Hyo Jeong Academic Foundation, Korea

Dr. Frank Kaufmann
President,
Professors World Peace Academy, USA

Dr. Jin Choon Kim
Co-Chair, Cheon Il Guk Academy
of Arts and Sciences, Korea

Dr. Theodore Tadaaki Shimmyo
Preesident Emeritus,
Unification Theological Seminary

Mr. Glenn Strait
Former Senior Editor,
World & I Innovative Approaches to Peace, USA

Dr. Frederick Swarts
Life Sciences Editor,
New World Encyclopedia,, USA

Dr. Jong Choon Woo
Professor Emeritus, Department of Forestry,
Kangwon University, Korea

Master of Ceremonies

Dr. Tyler Hendricks, President Emeritus, Unification Theological Seminary, Korea

Staff

Dr. Jun Young Cha, Executive Director, HJIFUS-Korea

Mr. Gregg Jones, Outreach Director, HJIFUS, USA

Ms. Jimin Moon, Assistant Administrator, HJIFUS-Korea

Mr. Shimpuku Uezono, Associate Research Director, HJIFUS, USA

Mr. Ryu-Sung Weinmann, Associate Administrative Director, HJIFUS, USA

* We wish to thank the many additional staff members who helped as volunteers for ICSG III.

Attendees List of ICSG III

Amemiya, Yoshiyuki
Professor Emeritus,
The University of Tokyo, Japan
PhD, Applied Physics and Engineering

Aoki, Kazuhiro
Professor,
Tokyo Medical and Dental University, Japan
PhD, Dentistry, Pharmacology

Bahng, GunWoong
Chair, Session 3
Professor,
Department of Mechanical Engineering,
The State University of New York, Korea
PhD, Materials Science and Engineering

Bailey, Patrick G.
Commentator, Session 2
President, Institute for New Energy, USA
PhD, Nuclear Engineering and Physics

Beauregard, Pierre
Canadian representative, IMAP, Canada
Master's in Religious Education

Berg, Staffan
IAAP-US, USA
DMin

Bulow, Nancy
Former Executive Assistant,
World Media Association, USA
MA, Religious Education

Cha, Jun Young
Member, Board of Directors, HJIFUS-Korea
Executive Director, HJIFUS-Korea
PhD, Literature

Choe, Joonghyun
Former Professor, Sun Moon University, Korea
PhD, Religious Studies

Choi, Byunghwan
Honorary Professor, Daejeon University, Korea
PhD, Philosophy

Choi, Jung Chang
Former President, The Research Institute
for the Integration of World Thought, USA
PhD, Ministry

Davis, Thomas
Professor Emeritus,
University of New Hampshire, USA
PhD, Genetics

Delorme, Arnaud
Commentator, Session 1
Associate Researcher,
Institute for Neural Computation,
University of California, San Diego, USA
PhD, Computational Neuroscience

Filmore, Paul
Chair, Session 2
Visiting Senior Research Fellow, Computing
and Mathematics, University of Plymouth, UK
PhD, Physics

Guignard, Didier
President, Club Science Paris, France
MSc, Electrical Engineering

Han, Chul Hee
Member, Board of Directors, HJIFUS-Korea
Professor, Chemistry,
Sun Moon University, Korea
PhD, Physical Chemistry

Hayashi, Masahisa
Professor Emeritus, Waseda University;
Representative Director,
Institute for Peace Policies, Japan
PhD, Public Finance

Hendricks, Tyler
Master of Ceremonies
Theological Consultant,
Family Federation for World Peace and
Unification International, Korea
PhD, Religion

Hiroi, Takahiro
Senior Research Scientist,
Brown University, USA
PhD, Mineralogy

Holden, Janice
Commentator, Session 1
Professor Emerita, Counseling Program,
University of North Texas, USA
Doctor of Education

Ishii, Hiroshi
Commentator, Session 2
Director, Kawasaki Kokoro Hospital, Japan
MD, PhD, Geriatric Neuropsychology

Ito, Shoichi
Professor Emeritus, Kyushu University, Japan
PhD, Food Economist

Jin, Sung Bae
Member, Board of Directors, HJIFUS-Korea
Chairman, HyoJeong Academic Foundation
Korea
PhD, Philosophy of Science

Joo, Douglas D.M.
Chair, ICSG III and Organizing Committee
Chair, Board of Directors; President, HJIFUS
Dr., Political Science

Joo, Jaewan
Professor, Sun Moon University, Korea
PhD, Theology

Kafatos, Menas
Chair, Session 1
Fletcher Jones Endowed Professor
of Computational Physics, and Director,
Institute for Earth, Computing, Human and
Observing, Chapman University, USA
PhD, Physics

Kato, Shinyu
Graduate Student, The University
of Electro-Communications, Japan

Kato, Takashi
Professor, Graduate School,
Chiba University, Japan
PhD, Theology, Biblical Studies

Kato, Tetsuji
Graduate Student,
The University of Tokyo, Japan

Kim, Hi Seong
Graduate Student,
Seoul National University, Korea

Kim, Jin Choon
Member, Board of Directors, HJIFUS-Korea
Honorary Professor, SunHak Universal Peace
Graduate University, Korea
PhD, Physics

Kim, Won H.
Commentator, Session 2
Professor, Biochemistry,
Wonju College of Medicine,
Yonsei University, Korea
PhD, Biochemistry

Kojima, Koki
Researcher,
Global Human Resources Agency, Korea
PhD, Theology; MA, Computer Security

Kuribayashi, Minoru
Associate Professor, Okayama University, Japan
PhD, Communication Engineering

Laszlo, Chris
Presenter, Session 3
Professor, Organizational Behavior,
Case Western Reserve University, USA
PhD, Economics

Lee, Jae Young
Chair-professor, Sun Moon University, Korea
Chair-professor,
Seoul University of Buddhism, Korea
PhD, Religious Studies

Lee, Sangmin
Researcher, Russia National Academy of Sciences Oriental Research Institute, Korea
PhD, International Relations

Ludwig, Thorsten
Commentator, Session 2
Senior Scientist and Owner,
Dr. Ludwig New Energy Technologies, Germany
Dr.rer.nat.

Miller, Lisa Jane
Keynote Speaker
Professor, Clinical Psychology,
Columbia University, USA
PhD, Psychology

Minami, Katsuyuki
Professor Emeritus,
Soil Science and Environmental Science,
Kitasato University, Japan
PhD, Agriculture

Moffitt, Larry
Member, Board of Directors, HJIFUS, USA
Executive Director,
The Washington Times Foundation, USA
MSc, Communication

Moon, Yeon Ah
Member, Board of Directors,
HJIFUS, USA; HJIFUS-Korea
Chairman,
Universal Peace Federation Korea
PhD, Sociology

Oh, Hungkuk
Honorary Professor, Aju University, Korea
PhD, Mechanical Engineering

Ohkado, Masayuki
Commentator, Session 1
Professor, Graduate School of Global Humanics,
Chubu University, Japan
PhD, Liberal Arts

Oizumi, Hiroko
Former Member,
House of Representatives, Japan
Master of Administrative Science

Okamoto, Satoshi
Professor, Chubu University, Japan
PhD, Literature

Park, Dongsun
Professor, Pusan National University, Korea
PhD, International Relations

Park, Koonam
Lab Manager/Research Associate,
Dermatology at School of Medicine,
Yale University, USA
MSc, Molecular Biology, Immunology

Park, Richard
Director, NuclixBio, Inc., Korea
PhD, Biology

Powell, Diane Hennacy
Commentator, Session 1
Neuropsychiatrist; Independent Researcher;
Treasurer, Parapsychological Association, USA

Pyun, Young Sik
Honorary Professor,
Sun Moon University, Korea
PhD, Industrial Engineering
and Nano Technology

Rashed, Fatma
Researcher,
Tokyo Medical and Dental University, Japan
PhD, Oral Biology

Rein, Glen
Commentator, Session 3
Founder and CEO,
Quantum-Biology Research, USA
PhD, Neurochemistry

Ross, Christina
Presenter, Session 2
Biophysicist, Wake Forest School of Medicine,
Winston-Salem, USA
PhD, Energy Medicine

Rossi, Giuseppe
Former Contract Professor,
University of Catanzaro, Italy
Holistic Doctor; Acupuncturist
MD, Worker's Diseases

Rubik, Beverly
Commentator, Session 3
Professor, Energy Medicine University, USA
PhD, Biophysics

Sa, Kwang Kee
Co-chair, Cheon Il Guk Academy
of Art and Sciences, Korea

Sakamaki, Yoshika
Trainer of Students,
SunHak Peace Academy, Korea
MA, Pastoral Studies

Schwartz, Gary
Presenter, Session 3
Professor, Psychology, Medicine,
Neurology, Psychiatry, and Surgery,
University of Arizona, USA
PhD, Psychology

Selover, Thomas
President,
SunHak UP Graduate University, Korea
ThD, Comparative Religion

Shimmyo, Theodore
Member, Board of Directors, HJIFUS, USA
President Emeritus,
Unification Theological Seminary, USA
PhD, Theology

Strait, Glenn
Member, Board of Directors, HJIFUS, USA
Former Senior Editor, The World & I: Innovative
Approaches to Peace, USA

Storberg-Walker, Julia
Commentator, Session 3
Department Chair,
Human and Organizational Learning,
The George Washington University, USA
PhD, Work, Community, and Family Education

Takahashi, Kazuo
Commentator, Session 3
Former Professor,
International Christian University, Japan
PhD, Political Science

Tanimoto, Shoichi
Researcher,
Institute for Molecular Science, Japan
PhD, Theoretical Chemistry

Terasawa, Kunihiko
Associate Professor, Wartburg College, USA
PhD, Comparative Religions and Interreligious
Dialogue; Buddhist-Christian Dialogue

Togami, Yoshihiro
Professor,
Tezukayama Gakuin University, Japan
PhD, Electrical and Electronic Engineering

Valone, Thomas F.
Presenter, Session 2
President, Integrity Research Institute, MD, USA
PhD, General Engineering

Wahbeh, Helané
Presenter, Session 1
Director of Research, Institute of Noetic Sciences, USA
ND, MCR, Naturopathic Medicine, Clinical Research

Woo, Jong Choon
Member, Board of Directors, HJIFUS-Korea
Professor Emeritus, Kangwon National University, Korea
PhD, Forestry

Woollacott, Marjorie
Presenter, Session 1
Professor Emeritus, Institute of Neuroscience, University of Oregon, USA
PhD, Cell and Molecular Biology

Yang, Susan
Professor of Computational Neuroscience, Chapman University, USA
PhD, Computational Neuroscience

Founders of ICSG and HJIFUS

The Reverend Dr. Sun Myung Moon and the Reverend Dr. Hak Ja Han Moon began life in Holy Marriage in 1960. They based their life on the understanding that the family is the school of love and that all of humanity is one extended family, with God as our common Parent. They wholeheartedly dedicated themselves to practicing their teaching of true love and working to create a peaceful world. Faithful to this divine calling, they came to embody the philosophy of living for the sake of others and their mission as True Parents of humankind. They found that joyfully living for others resembles God's investment in the creation of the world and human beings. Living for the sake of others is the secret to resolving conflict and ending suffering in the world.

ICUS IV, 1975: Rev. Dr. Sun Myung Moon and Rev. Dr. Hak Ja Han Moon welcome participants in New York City.

Working side by side for more than 50 years, they founded and developed a vast array of organizations devoted to establishing peace, including the Holy Spirit Association for the Unification of World Christianity (HSA-UWC), Family Federation for World Peace and Unification

ICUS VIII, 1979: Rev. Dr. Moon delivers his Founder's Address at the Opening Plenary Session in Los Angeles.

(FFWPU), the Women's Federation for World Peace (WFWP), the Youth Federation for World Peace (YFWP), the Professors World Peace Academy (PWPA), and the Universal Peace Federation (UPF). They founded prominent media organizations such as *The Washington Times* and *The Segye Times,* among many others around the world. In addition, they established international institutions that support education, industry, sports, the arts, and humanitarian projects. Guided by principles that transcend differences between races, religions, and cultures, these organizations have been working to realize a world of interdependence, mutual prosperity, and universally shared values.

During the Cold War Era, the founders emphasized the need to resolve three core arenas of disorder in the world: communism, religious conflict, and the breakdown of the family. Based on their first-hand experiences in war-torn Korea, they saw that the militant atheism and class warfare advocated by communist forces had led to unspeakable atrocities and suffering of millions. In response, the founders built ideological and educational insti-

ICUS XIV, 1985: In front (*left to right*): Alvin Weinberg, Marcelo Alonso, Durwood Foster, Kenneth Mellanby, Sun Myung Moon, Hak Ja Han Moon, Morton A. Kaplan, Eugene Wigner, Chung Hwan Kwak, and Bo Hi Pak. In rear (*left to right*): W.W. Bartley, III, Tor Ragnar Gerholm, Huston Smith, Claude Villee, Alexander King, John Oxenham, Vicenzo Cappelletti, Jean Charon, and Neil A. Salonen.

tutions, such as the International Federation for Victory Over Communism (IFVOC), that challenged these forces and offered counterproposals that contributed significantly toward ending the Cold War.

They tackled head-on the dark history of religious conflicts through the International Religious Foundation (IRF), the American Clergy Leadership Conference (ACLC) and UPF's Middle East Peace Initiative (MEPI). By their example and sacrifice, they pioneered an ever-expanding federation of religious leaders working together in all parts of the world to foster a new era of religious harmony among believers from all faiths.

They taught that science is a necessary component in the pursuit of human happiness and progress. In 1972, they founded the International Conference on the Unity of the Sciences (ICUS). Based on their vision of interdisciplinary cooperation, ICUS conferences have brought together scien-

ICUS XXIII, 2017: Rev. Dr. Hak Ja Han Moon (*center*) speaks to ICUS XXIII participants about the responsibility of scientists to protect nature. Sitting with her are Dr. Sun Jin Moon (*left*), Chair of the Pacific Rim Education Foundation, and Dr. Yeon Ah Moon (*right*), Chair of Universal Peace Federation, Korea

tists and scholars from the natural sciences, social sciences, and the humanities to work collaboratively with a holistic perspective pursuing the unity of the sciences with absolute values—that is, universal values and the fundamental cause and nature of the universe.

After the ascension of Sun Myung Moon in 2012, his wife, Hak Ja Han Moon, took responsibility for these projects with resolve. In honor of her husband's life, she established the Sunhak Peace Prize Foundation to recognize and award righteous heroes of our time who have taken the lead in the aspects of sustainable human development, conflict resolution, and ecological preservation. She introduced many new organizations such as the Hyo Jeong World Peace Foundation (HJWPF) and the World Christian Leadership Conference (WCLC) to expand on previous activities. *Hyo Jeong* means "filial heart"—indicating the teaching of a heart of filial piety toward God. She expanded UPF with initiatives like the International Association of Parliamentarians for Peace (IAPP), the International Summit Council for Peace (ISCP), the Interreligious Association for Peace and Development (IAPD), the International Association of Academicians for Peace (IAAP),

At the **Southeast Europe Peace Summit** organized by UPF, eight former heads of state from nations that had been in conflict during the Balkan wars receive ISCP founding medals from Rev. Dr. Hak Ja Han Moon (*center*), October 27, 2019, Tirana, Albania.

the International Media Association for Peace (IMAP), the International Association for Peace and Economic Development (IAED), and the Think Tank project which particularly gathers the wisdom and expertise of leaders in all sectors of society around the world who want to see the peaceful unification of the Korean Peninsula.

Among these new organizations, the Hyo Jeong International Foundation for the Unity of the Sciences (HJIFUS) was founded with the mission of solving environmental problems and living in harmony with nature. Rev. Dr. Hak Ja Han Moon has encouraged scientists and academics to pool their knowledge together to effectively address these issues. She has advocated the development of water purification systems for use in Africa and solving fine dust and air pollution in Eastern Asia. She expressed hope that the people of China, using modern technology, might transform the Gobi Desert into a verdant forest. The founders have expressed that to fulfill the mission of environmental protection and restoration, we need to clearly understand that the cosmos is like one body in which we are inseparably connected to the natural world and its fundamental cause. ◊

Introduction to HJIFUS

The Hyo Jeong International Foundation for the Unity of the Sciences (HJIFUS) is a nonprofit organization with the mission to protect and restore the Earth's natural environment. It was founded on the inspiration of Rev. Dr. Sun Myung Moon and Rev. Dr. Hak Ja Han Moon, who have elucidated a unique understanding of the relationships between God, humans, and nature.

Our current environmental problems include climate change; pollution of air, land, and water; changes in ecosystems; challenges in generating clean energy; and depletion of natural resources. These issues have detrimental consequences, such as deterioration in the quality of food and adverse effects on the health of humans and other species.

Modern science and technology are powerful tools that shape our world. When used competently and with the right intentions, they benefit both humans and nature. However, when misused through human selfishness, greed, or ignorance, they cause widespread environmental damage.

To help overcome environmental problems stemming from human behavior, HJIFUS promotes character education to improve the human attitude toward nature. On a practical level, the foundation identifies and helps implement well-rounded technological solutions and methods derived from both conventional science and frontier science. In turn, these technologies and how to use them are freely shared among all levels of society regardless of race, nationality, religion, social status, or culture. HJIFUS envisions a world where human beings live in transparency as one global family, based on the ideals of interdependence, mutual prosperity, and universally shared values. It is in this way that environmental crises can ultimately be surmounted.

ICUS XXIII participants with HJIFUS founder, Rev. Dr. Hak Ja Han Moon (*front row, center*), February 2017, Seoul, Korea.

To accomplish these goals, HJIFUS works in six areas. One area of HJIFUS activity is the hosting of international environmental conferences. The International Conference on the Unity of the Sciences (ICUS) was originally started by the founders in 1972 to bring scientists and academics together pursuing the unity of the sciences with absolute values. The ICUS conferences were held regularly until the year 2000. From 2017, HJIFUS reconvened ICUS and has been hosting these conferences annually with a new focus to resolve environmental challenges. These meetings examine technologies derived from conventional science. In addition, HJIFUS has launched another conference series, beginning with the First International Conference on Science and God (ICSG I) in 2020. This series considers environmental solutions offered by frontier science, including postmaterialist science, which may represent a new dawn in the history of science.

In a second area, HJIFUS develops and offers environmental policy proposals to national and international agencies and policy makers. This is accomplished partly by distributing conference proceedings containing such

ICUS XXIV, February 2018, Seoul, Korea: Rev. Dr. Hak Ja Han Moon (*right*) shares her vision with young scientists, encouraging them to lead future developments in protecting and restoring the Earth environment.

proposals. To broaden the scope of its activities and to collect consensus from experts and professionals around the world, HJIFUS is considering to establish a World Environmental Council or Congress.

In a third area, HJIFUS produces periodicals and other publications useful for educating the general public on the environment. On Earth Day, April 22, 2021, HJIFUS launched *The Earth & I,* a bimonthly online magazine that provides the latest news on local and global environmental issues, innovative technological solutions, advice on health and life, interviews with leading experts, and human-interest stories and testimonies related to improving the environment. Future publication plans include producing an encyclopedia of environmental science and technology, a *Who's Who* of key people and organizations in the environmental arena, a map of the global environment, and overviews of environmental policies in countries around the world.

The fourth area of HJIFUS activity involves leading social campaigns and programs to promote environmental restoration through service projects, hands-on education, and experiencing nature first-hand so that people

HJIFUS representatives visit Senegal and collect water samples for testing, June 2019.

of all ages can develop a caring and respectful relationship with the natural world. In July 2021, HJIFUS formed an outreach department to coordinate these activities and to explore partnerships with like-minded organizations.

The fifth area HJIFUS pursues is to raise the living conditions of people through practical projects. From 2018, HJIFUS has been laying the groundwork for a project that supplies devices that produce healthy drinking water for communities in need in Africa and elsewhere around the world.

Finally, HJIFUS supports the manufacture of everyday products that can substitute for items that exacerbate environmental problems. Through such products, people can live greener lifestyles and improve their health and well-being while taking care of the Earth.

With these approaches and multipronged strategies, HJIFUS is positioned to serve the international community with carefully thought-out plans, innovative technologies, and proven methods to protect, restore, and cherish the Earth for ages to come. HJIFUS organizations have been established in different parts of the world. They operate independently but share the same vision and mission. ◇

Boards of Directors

Supporting Organizations and Worldwide HJIFUS Offices

Sponsoring Organizations

Hyo Jeong International Foundation for the Unity of the Sciences–Korea
Hyo Jeong International Foundation for the Unity of the Sciences, USA

Supporting Organizations

The Washington Times Foundation
The Segye Times
Universal Peace Federation International
Family Federation for World Peace and Unification-International

Worldwide HJIFUS Offices

Korea
1st floor, 324-211 Misari-ro, Seorak-myeon Gapyeong-gun, Gyeonggi-do
Republic of Korea
Tel: +82-(0)70-7771-2604
E-mail: admin.kr@icus.org

United States
3600 New York Avenue, NE
Washington, DC 20002
USA
Tel: +1-202-636-2874
E-mail: admin@hjifus.org

Communication Offices

Japan
2nd Fl., Seiyaku Bldg.
5-13-2, Shinjuku, Shinjuku-ku
Tokyo 160-0022
Japan
Tel: +81-3-5362-0630
E-mail: world@pwpa-j.net

Europe
Espace Barrault
98 rue Barrault
75013 Paris
France
Tel: +33-6-68-31-63-87
E-mail: guignarddidier@gmail.com

Index

A

B

C

D

E

L

M

N

O

P

Q

R

S

T

U

V

W

XYZ

Made in United States
North Haven, CT
11 November 2026

22113770R00141